Annals of Mathematics Studies

Number 31

ANNALS OF MATHEMATICS STUDIES

Edited by Emil Artin and Marston Morse

1. Algebraic Theory of Numbers, *by* HERMANN WEYL
3. Consistency of the Continuum Hypothesis, *by* KURT GÖDEL
6. The Calculi of Lambda-Conversion, *by* ALONZO CHURCH
7. Finite Dimensional Vector Spaces, *by* PAUL R. HALMOS
10. Topics in Topology, *by* SOLOMON LEFSCHETZ
11. Introduction to Nonlinear Mechanics, *by* N. KRYLOFF *and* N. BOGOLIUBOFF
14. Lectures on Differential Equations, *by* SOLOMON LEFSCHETZ
15. Topological Methods in the Theory of Functions of a Complex Variable, *by* MARSTON MORSE
16. Transcendental Numbers, *by* CARL LUDWIG SIEGEL
17. Problème Général de la Stabilité du Mouvement, *by* M. A. LIAPOUNOFF
19. Fourier Transforms, *by* S. BOCHNER *and* K. CHANDRASEKHARAN
20. Contributions to the Theory of Nonlinear Oscillations, Vol. I, *edited by* S. LEFSCHETZ
21. Functional Operators, Vol. I, *by* JOHN VON NEUMANN
22. Functional Operators, Vol. II, *by* JOHN VON NEUMANN
23. Existence Theorems in Partial Differential Equations, *by* DOROTHY L. BERNSTEIN
24. Contributions to the Theory of Games, Vol. I, *edited by* H. W. KUHN *and* A. W. TUCKER
25. Contributions to Fourier Analysis, *by* A. ZYGMUND, W. TRANSUE, M. MORSE, A. P. CALDERON, *and* S. BOCHNER
26. A Theory of Cross-Spaces, *by* ROBERT SCHATTEN
27. Isoperimetric Inequalities in Mathematical Physics, *by* G. POLYA *and* G. SZEGO
28. Contributions to the Theory of Games, Vol. II, *edited by* H. KUHN *and* A. W. TUCKER
29. Contributions to the Theory of Nonlinear Oscillations, Vol. II, *edited by* S. LEFSCHETZ
30. Contributions to the Theory of Riemann Surfaces, *edited by* L. AHLFORS *et al.*
31. Order-Preserving Maps and Integration Processes, *by* EDWARD J. MCSHANE
32. Curvature and Betti Numbers, *by* K. YANO *and* S. BOCHNER

ORDER-PRESERVING MAPS AND INTEGRATION PROCESSES

By EDWARD J. McSHANE

Princeton, New Jersey
Princeton University Press
1953

London: Geoffrey Cumberlege, Oxford University Press
L. C. Card 53-6583

Printed in the United States of America

CONTENTS

ORDER-PRESERVING MAPS AND INTEGRATION PROCESSES

INTRODUCTION

The various definitions of the Lebesgue integral and its generalizations may be classified, if we wish, into two subsets, one making essential use of some kind of norm or modulus, and another, less numerous, in which the central role is played by some order relation. For example, Bochner's integral [Bochner 1]* of a function whose values lie on a Banach space involves formation of successive approximations which converge in the sense of the norm in the Banach space. Again, in Stone's treatment [Stone 1] of the integral, order properties are first used to define a norm in a function space, and then the integral is defined by means of a convergence according to this norm. On the other hand, the definition of the Riemann integral by use of the Darboux upper and lower integrals rests on order properties, and so does Daniell's definition [Daniell 1] of integral, generalizing the Lebesgue and Radon (or Lebesgue - Stieltjes) integrals. The Perron integral, too, is defined as the function which is simultaneously the lower bound of a set of overestimates and the upper bound of a set of underestimates, and thus is one of those making essential use of order.

Daniell develops the integral by assuming that a subset E of the lattice F of all real-valued functions on a set T is mapped into the set of real numbers by an order-preserving mapping I_0 satisfying certain requirements. He then shows how this mapping can be extended, giving a mapping I whose domain contains E and whose range is contained in the reals; this mapping is order-preserving, and also has the closure and continuity properties expressed in the well-known convergence theorems of the Lebesgue theory. Thus Daniell's theory can be regarded as a study of a special case of the following problem. We are given two partially ordered sets F and G, and an order-preserving mapping I_0 of a subset E of F into G. We seek conditions on E, I_0, F, and G that will enable us to extend the domain of definition of I_0, producing say a mapping I of F_0 into G, in such a way that the enlarged domain F_0 shall have some useful closure properties and the extended mapping I shall have some useful kind of continuity property on its domain. It is to the study of this problem that the following pages are devoted. Because it is

*References are to the brief bibliography at the end of the study.

an exploitation of the order properties almost exclusively, it is to be expected that the territory which it more or less naturally takes in is in part different from that covered by the integration theories based largely on norms. If it gracefully furnishes interesting applications, its existence is justified. We believe that the examples exhibited in the last chapter will show this to be the case.

For the sake of simplicity we restrict the problem by assuming that F is a lattice. This is not a great restriction, since every partially ordered set F can be embedded in a lattice, and if the process of extension yields a domain containing points of that lattice not in the original F we have the privilege of ignoring them. In fact, in all but one of the applications in the last chapter F is isomorphic with the lattice of extended-real-valued functions on some domain. The situation with regard to G is different. For example, in the application in Section 29 the values of I_0 lie in the set G of bounded hermitian operators on a Hilbert space. These may be regarded as defining quadratic functions on Hilbert space, and thus embedded in the lattice of real functions on the Hilbert space. But the chief importance of the extended mapping I is that its values also are bounded Hermitian operators. If we knew only that the values corresponded to some real function on the Hilbert space, the result would have been without interest. Therefore we refrain from assuming that the image space G is a lattice.

This in fact constitutes one of the chief differences between the present treatment of integration and others that have preceded it. Thus, for example, in the fundamental paper of H. Freudenthal [Freudenthal 1] there is developed a type of Riemann-Stieltjes integral with values in a "partially ordered module," which is not merely a partially ordered linear system as we here define the term, but is a lattice. Likewise, M. H. Stone [Stone 3] and Hidegorô Nakano [Nakano 1, 2] define integrals whose values lie in a lattice. The difference, insofar as it concerns our present aims, is not trivial. For example, a (suitably closed) commutative algebra of hermitian operators is actually a lattice, but this is by no means superficially evident; while on the other hand it is obvious that this algebra is partially ordered.

To the best of my knowledge, there are two important publications on integration processes involving partially ordered spaces which are not lattices. These are [Bochner 1] and [Bochner and Ky Fan 1]. In both of these a Riemann-Stieltjes type of integral is used, in the former to extend the Bernstein-Widder theorem, and in the latter to obtain a representation of the general distributive order-preserving mapping of functions continuous on an interval (or circumference) into a partially ordered space. The principal result of [Bochner 1] is very close to the application here in Section 30.

One result of working with partially ordered sets instead of lattices is that we must first develop an adequate theory of closure, completeness, convergence and continuity. This we do in the first chapter, a trifle more extensively than is absolutely essential for later use. It is our strong suspicion that some interesting portions of lattice theory can be usefully generalized to less restricted partially ordered sets.

Our methods are of course related to those of Daniell. Some rather drastic changes necessarily result from replacement of the lattice of real-valued functions on a set T by the general lattice F, and the replacement of the real number system by the partially ordered set G. But in addition to these there are a couple of changes that would affect the discussion even of the case considered by Daniell. One of these concerns the use of auxiliary partial orderings. The set F is partially ordered by the relation $\geq$ in terms of which it is a lattice; but there may also be other partial orderings useful as auxiliaries. For example, the real-valued functions on an interval J are partially ordered by the usual relation $\geq$. But we can also partially order them by the relation $\gg$, where by $f \gg g$ we mean that to each point x^* of J there corresponds a neighborhood on which the supremum of g does not exceed the infimum of f. When we develop the Lebesgue-Stieltjes integral by extending the elementary integral on step-functions, we use systems of step-functions directed by $\gg$ or by its reverse $\ll$, and thereby avoid the sometimes tedious discussion of the phenomena at the boundaries of the intervals of constancy of the step-functions. This same device is also of service in the study of the Riemann-Stieltjes integral, as will be shown in a joint paper with T. A. Botts [McShane and Botts 1]. With other special meanings for $\gg$, the auxiliary partial ordering also proves useful in the other examples in this paper.

The other departure from Daniell's pattern is a relaxation of the requirement that the set E of elementary functions on which the elementary integral I_0 is defined should be a lattice. This relaxation can be pictured with the help of the family of polynomials on an interval J. Since the (real) functions continuous on J form a lattice, for any three such functions g_1, g_2, g_3 the function $\text{mid}(g_1, g_2, g_3)$ whose value at each point is the middle one of the values of g_1, g_2, g_3 is also continuous. But if e_1, e_2, and e_3 are polynomials their "middle" is not necessarily a polynomial. However, in this case, if f is any function such that $f - \text{mid}(e_1, e_2, e_3)$ has a positive lower bound, by Weierstrass' theorem we can find a polynomial e on J such that $f - e$ and $e - \text{mid}(e_1, e_2, e_3)$ both have positive lower bounds on J. Our substitute for the requirement that E be a lattice in F is just the analogue of demanding that such an element e exist (and the dual requirement, where $\text{mid}(e_1, e_2, e_3) - f$ has a positive lower bound). Even this we ask only when $I_0(e_1)$ and $I_0(e_2)$

have a common upper bound. This relaxation is useful, for example, in studying the spectral resolution of bounded hermitian operators on a Hilbert space, as discussed in Section 29; for here the natural starting point is the family of polynomials in the operator. It is also useful in the applications studied in Sections 30, 31 and 32, for similar reasons. Because the requirement that there exist an e as just described is imposed only when $I_0(e_1)$ and $I_0(e_2)$ have a common bound, the integral obtained is not necessarily absolutely convergent. In fact, as one application of the general theory we obtain (in Section 27) an integral which closely resembles the Perron integral, and may be identical with it.

Our principal type of convergence based on order properties shows a close affinity with "pointwise" rather than "uniform" convergence. For instance, in Section 3 we show that when F is the set of extended-real-valued formations on some domain, order-convergence is the same as pointwise convergence. When F is the set of bounded hermitian operators on Hilbert space, order-convergence coincides with (eventually) bounded strong convergence. Accordingly, the spectral resolution of an operator B which we obtain serves to put the Borel-measurable functions f on the interval $[-||B||,||B||]$ into correspondence with operators, designated as usual by $f(B)$, forming a commutative algebra; the integral composes each $f(B)$ out of limits of linear combinations of certain projection-operators forming the resolution of the identity for B. When we consider algebras of hermitian operators (as in Section 31), we map the elements of the algebra on Borel-measurable functions on a product of certain intervals; this is in contrast with the Gelfand-Neumark representation by means of simpler (i.e., continuous) functions on a more complicated domain. For instance, if B is a bounded hermitian operator and E_λ, $-\infty < \lambda < \infty$ is its resolution of the identity, the least algebra containing B and all the E_λ is by the present theory represented by the Borel-measurable functions on the interval $[-||B||,||B||]$. The representation by continuous functions on a domain D would require D to be of much greater complexity.

These considerations may lend some interest to the theorems in Sections 31 and 32 concerning partially ordered algebras, since we again obtain theorems which, when interpreted in the specialization to operators on Hilbert space, give the "strong" rather than the "uniform" theory.

The examples which we have considered do not exhaust the possibilities, nor have the individual examples been studied as exhaustively as possible. It is hardly to be expected that the present approach would yield startlingly new results in such well-worked fields as Lebesgue-Stieltjes integration or the spectral resolution of a bounded hermitian operator. In familiar settings we are content to obtain familiar theorems; the novelty, if any, lies in the unification, whereby such apparently diverse processes as Perron-type integration and spectral resolution appear as instances subsumed under a single theory.

CHAPTER I

PARTIALLY ORDERED SETS AND SYSTEMS

§1. PARTIAL ORDERINGS AND LATTICES

A binary relation $>$ between pairs of elements of a set F is a <u>partial</u> <u>ordering</u> of F if it satisfies the conditions

(1.1) (a) If $a > b$ and $b > c$ then $a > c$.
(b) If $a > b$ and $b > a$ then $a = b$.

(Sometimes (1.1a) alone is used as the definition of a partial ordering; then relations satisfying (1.1a,b) are called <u>proper</u> partial orderings. But we shall adhere to the terminology above.) If $>$ is a partial ordering of F, and $a \geqq b$ is defined to mean that $a > b$ or $a = b$, we see that $\geqq$ is another partial ordering. It is reflexive ($a \geqq a$ for all a in F), and is the same as $>$ if $>$ was itself reflexive. Thus any partial ordering whose symbol includes the mark $=$ (as in $\geqq$) is reflexive, but the absence of $=$ in the symbol does not imply the absence of reflexivity.

Let F be partially ordered by $>$. As usual, $a < b$ shall mean $b > a$. We verify at once that $<$ is also a partial ordering of F. An element b of F is an upper bound for a subset S of F if $b \geqq x$ for all x in S; it is the <u>supremum</u> of S if it is an upper bound for S and moreover every upper bound b' of S satisfies the relation $b' \geqq b$. If b and b' are both suprema of S, both relations $b \geqq b'$ and $b' \geqq b$ hold, so $b' = b$. Thus the supremum of S, when it exists, is unique. It will be designated by the symbol $\bigvee S$. Lower bounds and the infimum of S are defined dually; the infimum of S, when it exists, will be designated by the symbol $\bigwedge S$. When S has just two members, say $S = \{a,b\}$, we usually write $a \vee b$ for $\bigvee S$ and $a \wedge b$ for $\bigwedge S$.

If a set F is partially ordered by $\geqq$, so is each subset F_o. But the situation here is not as simple as the analogous one in topology. For if S is a subset of F_o, it may have a supremum when regarded as a subset of F_o but not when regarded as a subset of F, or vice versa; or it may have a supremum in F_o when regarded as a subset of F_o and a different supremum in F when regarded as a subset of F. (Examples: (1) F the rationals, F_o all rationals x except those such that $2 < x^2 < 4$, S the rationals x such that $x^2 < 2$. (2) F the reals, F_o the rationals, S as before. (3) F the reals, F_o all reals x such that $x < 0$ or $x \geqq 1$, S the negative reals.) Consequently, when we are discussing partially ordered sets and their subsets, we must indicate clearly when the subset F_o is being treated as a subset of F, so that such symbols $\bigvee S$ shall mean that member of F which is the supremum of S, regarded as a subset of F. Usually we shall use the word "embedded" in definitions of concepts in which this subset relationship is essential. Thus in a definition or theorem involving a set F_o "embedded" in F, whenever S is a subset of F_o the symbol $\bigvee S$ shall always mean that member of F which is the supremum of S when S is considered as a subset of F, without reference to F_o.

A set F_o contained in a partially ordered set F is a <u>lattice embedded in</u> F if for each a and b in F_o the supremum $a \vee b$ and the infimum $a \wedge b$ exist (meaning that they exist in F) and are themselves members of F_o. When F_o can be chosen to be F itself, instead of saying "F is a lattice embedded in F" we say for brevity "F is a lattice." When F_o is a lattice in F, by successive applications of the operations $\vee$ and $\wedge$ we form "lattice combinations" of elements of F_o. These may be defined inductively as follows. The identity function on F_o, whose value at f is f, is a "lattice combination." For $n > 1$, a function L on ordered n-uples $f_1, \dots, f_n$ of elements of F_o is a lattice combination if we can find an integer m $(1 \leqq m < n)$, a lattice combination L_1 of $f_1, \dots, f_m$ and a lattice combination L_2 of $f_{m+1}, \dots, f_n$ such that L has one of the two forms $L_1 \vee L_2$, $L_1 \wedge L_2$. It is easily seen that every lattice combination of elements of a lattice F_o is again a member of F_o. Also, $\bigvee\{a,b,c\} = a \vee (b \vee c) = (a \vee b) \vee c$, and likewise for all finite subsets of F_o.

A set F is <u>directed</u> by a partial ordering $>$ if to each pair a,b of elements of F corresponds an element c of F such that $c > a$ and $c > b$. Evidently, if F is a lattice under the partial ordering $\geqq$, it is directed both by $\geqq$ and by $\leqq$. The following lemma expresses a simple property of lattices which we shall utilize several times.

(1.2) LEMMA. If F is a lattice embedded in a partially ordered set F', and S is a non-empty subset of F, and S* is the set of all suprema of finite subsets of S, and either $\bigvee S$ or $\bigvee S^*$ exists, then both $\bigvee S$ and $\bigvee S^*$ exist, and they are equal.

If b is an upper bound for S* it is an upper bound for S,

which is contained in S^*. If b is an upper bound for S it is also an upper bound for each finite subset of S, hence $b \geqq s^*$ for each s^* in S^*, and b is an upper bound for S^*. So each upper bound for either of S, S^* is an upper bound for the other, whence the conclusion follows.

Even in the case of real numbers there is great advantage in considering, along with the ordering $\geqq$ the "stronger" ordering $>$. In the applications of the present theory such an idea of "strengthening" of a partial ordering proves quite useful. Accordingly we set down a definition:

(1.3) DEFINITION. If F is a set partially ordered by a relation $>$, and $\gg$ is a binary relation on F, we say that $\gg$ is a "strengthening" of $>$ if for all elements f,g,h,k of F the following statements hold.

(a) If $f \gg g$ then $f > g$.

(b) If $f \gg g$ and $g \geqq k$, or if $f \geqq g$ and $g \gg k$, then $f \gg k$.

(c) If $f \gg h$ and $f \gg k$, there exists an element e of F such that $f \gg e$, $e \geqq h$ and $e \geqq k$; and dually.

It is easy to see that if F is a lattice and (1.3a,b) are satisfied, (1.3c) is equivalent to the simpler statement

(1.3c') If $f \gg h$ and $g \gg k$, then $f \vee g \gg h \vee k$ and $f \wedge s \gg h \wedge k$.

§ 2. COMPLETENESS

In the theory of lattices a concept of completeness has been introduced (Birkhoff [1] p. 16) which generalizes the idea used by Dedekind in developing the real number system. It is equivalent to this.

(2.1) DEFINITION. A lattice L embedded in a partially ordered set F is σ-complete, if every countable non-empty subset of L has a supremum and an infimum (in F), and these both belong to L; conditionally σ-complete, if every countable non-empty subset of L which has a lower bound in L also has an infimum which also belongs to L, and every countable non-empty subset which has an upper bound in L also has a supremum which also belongs to L; complete, or conditionally complete, if the corresponding one of the above statements holds with the word "countable" omitted. (Our terminology differs somewhat from that of Birkhoff.)

This definition exhibits a phenomenon which will recur throughout the following chapters. The definitions and theorems often occur in pairs, one using an assumption of countability and the other not. In order to avoid verbosity, we shall adopt the following convention. In the statements of definitions, and in the statements and proofs of theorems, and occasionally even in their titles, certain expressions are bracketed. If we include all such bracketed expressions, we obtain one definition or theorem and proof; if we reject all the bracketed expressions, we obtain another definition or theorem and proof.

Clearly a set directed by $\geqq$ and by $\leqq$ could not be even conditionally σ-complete without being a lattice. Accordingly, for directed sets which are not necessarily lattices we need a different concept. Such a concept was defined by H. Freudenthal (Freudenthal [1]) and utilized also by M. H. Stone (Stone [3]) and by S. Bochner (Bochner [2]). We use the following form, somewhat different from but closely related to theirs:

(2.2) DEFINITION. A partially ordered set F is Dedekind $[\sigma\text{-}]$complete if for every [countable] non-empty subset S of F which is directed by $\geqq$ and has an upper bound in F, the supremum $\bigvee S$ exists in F, and for every [countable] non-empty subset S of F which is directed by $\leqq$ and has a lower bound in F, the infimum $\bigwedge S$ exists in F.

(2.3) COROLLARY. If F is a lattice, it is conditionally $[\sigma\text{-}]$complete if and only if it is Dedekind $[\sigma\text{-}]$ complete.

If F is conditionally $[\sigma\text{-}]$complete it is obviously Dedekind $[\sigma\text{-}]$complete. Suppose then that F is Dedekind $[\sigma\text{-}]$complete. Let S be a [countable] non-empty subset of F having an upper bound b^*. Let S^* be the set of all suprema of finite subsets of S. This set is [countable, and is] directed by $\geqq$; for if a, b are elements of S^*, the equations $a = \bigvee\{a_1, \ldots, a_n\}$, $b = \bigvee\{b_1, \ldots, b_m\}$ hold for proper choice of $a_1, \ldots, b_m$. Then the element $c = \bigvee\{a_1, \ldots, a_n, b_1, \ldots, b_m\}$ of S^* satisfies $c \geqq a$, $c \geqq b$. Hence by the Dedekind $[\sigma\text{-}]$completeness of F, $\bigvee S^*$ exists. By (1.2) this is $\bigvee S$. The existence of $\bigwedge S$ when S has a lower bound is proved dually.

A simple example of a partially ordered set (not a lattice) which is Dedekind complete but not conditionally complete is the set of all closed circular regions in the plane, ordered by inclusion.

The rather easy proof of the following lemma is omitted.

(2.4) **LEMMA.** If F is a partially ordered set and S is a countable subset of F directed by $\geqq$ and having a supremum, there exists a sequence $a_1, a_2, \ldots$ of elements of S such that $a_1 \leqq a_2 \leqq a_3 \leqq \ldots$ and $\bigvee_n a_n = \bigvee S$; and dually for countable sets directed by $\leqq$.

From this it is clear that

(2.5) COROLLARY. A partially ordered set F is Dedekind σ-complete if and only if every sequence of elements of F which increases with n and has an upper bound also has a supremum in F, and every sequence of elements of F which decreases with increasing n and has a lower bound also has an infimum in F.

This property equivalent to Dedekind σ-completeness is the one previously defined by Freudenthal.

It is customary to define the "cartesian product" of sets $X_\delta F_\delta$ of sets $\{F_\delta : \delta \text{ in } D\}$ to be the set of all functions $(\phi(\delta) : \delta \text{ in } D)$ such that $\phi(\delta)$ is in F_δ for each δ. If each set F_δ is partially ordered by a partial ordering $>_\delta$, we define the cartesian product $X_\delta[F_\delta, >_\delta]$ to be the system $[F, >]$ such that the elements of F are the functions $(\phi(\delta) : \delta \text{ in } D)$, and for two such elements ϕ_1 and ϕ_2, the relation $\phi_1 > \phi_2$ means that $\phi_1(\delta) >_\delta \phi_2(\delta)$ for all δ in D. It is obvious that $>$ is a partial ordering of F. Since we shall not have any use for cartesian products except when the sets are partially ordered, we shall usually abbreviate the product symbol to $X_\delta F_\delta$, the meaning of the partial ordering being understood as above. For ϕ in F, the "projection" of ϕ on F_δ shall mean $\phi(\delta)$; for a subset S of F, the "projection" of S on F_δ shall mean the set consisting of the projections on F_δ of the elements of S.

(2.6) **THEOREM.** Let $[F, >]$ be the cartesian product of partially ordered sets $[F_\delta, >_\delta]$, δ in D. Let S be a subset of F. In order that S have a supremum in F it is necessary and sufficient that for each δ in D, the projection of S on F_δ shall have a supremum (with respect to $>_\delta$); in this case, the supremum of S is the element of F whose projection on F_δ is the supremum of the projection of S on F_δ. The dual is also true.

If S has a supremum ϕ, for each δ the projection $\phi(\delta)$ is an upper bound for the projection of S in F_δ. Fix a δ' in D, and let b' be an upper bound for the projection of S in $F_{\delta'}$. Let $\phi'(\delta) = \phi(\delta)$ for δ in D, $\neq \delta'$, while $\phi'(\delta') = b'$. Then ϕ' is an upper bound for S, so $\phi' \geqq \phi$, and $b' \geqq_{\delta'} \phi(\delta')$. That is, $\phi(\delta')$ is the supremum of the projection of S on F_δ. Conversely, if $\phi(\delta) = \bigvee[\text{projection of } S \text{ on } F_\delta]$ for each δ, ϕ is an upper bound for S. Given any other upper bound ϕ' of S, we have for each δ in D that $\phi'(\delta) \geqq_\delta \psi(\delta)$ for all ψ in S, so $\phi'(\delta)$ is an upper bound for the projection of S on F_δ, and $\phi'(\delta) \geqq_\delta \phi(\delta)$. Hence ϕ is the supremum of S. The dual is established similarly.

(2.7) COROLLARY. If for each δ in D the set F_δ is Dedekind-complete under the partial ordering $<_\delta$, the cartesian product $\times_\delta[F_\delta, >_\delta]$ is also Dedekind complete.

If a lattice F does not contain a greatest element, we can adjoin a new ideal element (usually denoted by I) such that $I \geqq x$ for all x in F. Likewise if F does not contain a least element 0 we can adjoin one. If F is Dedekind complete or Dedekind σ-complete so is the enlarged lattice. More: each non-empty set $S \subset F$ has an upper bound I and a lower bound 0, so by (2.1) the enlarged lattice is a $[\sigma\text{-}]$complete lattice if F is Dedekind $[\sigma\text{-}]$complete.

If F is a complete lattice, and S_1 and S_2 are non-empty subsets of F, we readily prove

(2.8) $$(\bigvee S_1) \vee (\bigvee S_2) = \bigvee\{x \vee y : x \text{ in } S_1 \text{ and } y \text{ in } S_2\},$$
$$(\bigwedge S_1) \wedge (\bigwedge S_2) = \bigwedge\{x \wedge y : x \text{ in } S_1 \text{ and } y \text{ in } S_2\}.$$

The relations

(2.9) $$(\bigvee S_1) \wedge (\bigvee S_2) = \bigvee\{x \wedge y : x \text{ in } S_1 \text{ and } y \text{ in } S_2\},$$
$$(\bigwedge S_1) \vee (\bigwedge S_2) = \bigwedge\{x \vee y : x \text{ in } S_1 \text{ and } y \text{ in } S_2\}$$

may be true or false. If (2.9) holds for every pair S_1, S_2 of non-empty subsets of F, we say that F is "infinitely distributive." If (2.9) holds for finite sets, F is "distributive."

In a distributive lattice F we can establish the identity, valid for all a,b,c in F

(2.10) $$\bigwedge\{a \vee b,\ a \vee c,\ b \vee c\} = \bigvee\{(a \wedge b,\ a \wedge c),\ b \wedge c)\};$$

for

$$\bigvee\{a \wedge b,\ a \wedge c,\ b \wedge c\} = [a \wedge b] \bigvee\{(a \wedge c) \bigvee (b \wedge c)\}$$
$$= [a \bigvee\{(a \wedge c) \bigvee (b \wedge c)\}] \wedge [b \bigvee\{(a \wedge c) \bigvee (b \wedge c)\}]$$
$$= [a \bigvee (b \wedge c)] \wedge [b \bigvee (a \wedge c)]$$
$$= [(a \vee b) \wedge (a \vee c)] \wedge [(b \vee a) \wedge (b \vee c)]$$
$$= \bigwedge\{a \vee b,\ a \vee c,\ b \vee c\}$$

The quantity defined by either member of (2.10) is called the "middle" of a,b,c.

(2.11) DEFINITION. If F is a distributive lattice, for each three elements a,b,c of F the "middle" of a,b,c is defined to be

$$\mathrm{mid}(a,b,c) = \bigvee\{a \wedge b,\ a \wedge c,\ b \wedge c\}$$
$$= \bigwedge\{a \vee b,\ a \vee c,\ b \vee c\}$$

We shall need some of the properties of this combination.

(2.12) THEOREM. In a distributive lattice F, if $b \geqq a$ then

$$\mathrm{mid}(a,b,c) = b \wedge [a \vee c] = a \vee [b \wedge c] \ .$$

This is obvious from (2.11), since $a \vee b = b$ and $a \wedge b = a$.

(2.13) THEOREM. In a distributive lattice F,

$$\mathrm{mid}(x,y,z) = \mathrm{mid}\ (x \wedge y,\ x \vee y, z) \ .$$

In (2.12) replace a by $x \wedge y$, b by $x \vee y$, and c by z; then

$$\mathrm{mid}\ (x \wedge y,\ x \vee y, z) = (x \vee y) \wedge [(x \wedge y) \vee z]$$
$$= (x \vee y) \wedge [(x \vee z) \wedge (y \vee z)]$$
$$= \mathrm{mid}\ (x,y,z) \ .$$

(2.14) THEOREM. Let F be a distributive lattice, and let $\{a_\alpha : \alpha \text{ in } A\}$, $\{b_\beta : \beta \text{ in } B\}$ and $\{c_\gamma : \gamma \text{ in } \Gamma\}$

be finite subsets of F. Then

$$\bigvee_{\alpha,\beta,\gamma} \operatorname{mid}(a_\alpha, b_\beta, c_\gamma) = \operatorname{mid}(\bigvee_\alpha a_\alpha, \bigvee_\beta b_\beta, \bigvee_\gamma c_\gamma) .$$

If F is complete and infinitely distributive, this holds even when the three subsets are not required to be finite.

Consider first the special case in which there is only one a_α, which we call a, and only one b_β, which we call b. By (2.13), it is sufficient to establish the equation under the assumption $b \geqq a$. Then

$$\begin{aligned} \operatorname{mid}(a,b,\bigvee_\gamma c_\gamma) &= b \wedge (a \vee [\bigvee_\gamma c_\gamma]) \\ &= b \wedge (\bigvee_\gamma [a \vee c_\gamma]) \\ &= \bigvee_\gamma (b \wedge [a \vee c_\gamma]) \\ &= \bigvee_\gamma \operatorname{mid}(a,b,c_\gamma) . \end{aligned}$$

Returning to the general case,

$$\begin{aligned} \bigvee_{\alpha,\beta,\gamma} \operatorname{mid}(a_\alpha, b_\beta, c_\gamma) &= \bigvee_\alpha [\bigvee_\beta \{\bigvee_\gamma \operatorname{mid}(a_\alpha, b_\beta, c_\gamma)\}] \\ &= \bigvee_\alpha [\bigvee_\beta \operatorname{mid}(a_\alpha, b_\beta, \bigvee_\gamma c_\gamma)] \\ &= \bigvee_\alpha [\operatorname{mid}(a_\alpha, \bigvee_\beta b_\beta, \bigvee_\gamma c)] \\ &= \operatorname{mid}(\bigvee_\alpha a_\alpha, \bigvee_\beta b_\beta, \bigvee_\gamma c_\gamma) . \end{aligned}$$

§ 3. TYPES OF CONVERGENCE

Let ϕ be a function on a partially ordered set F to a partially ordered set G. We say that ϕ is isotone if for all pairs f,f' of elements of F, the relation $f' \geqq f$ implies $\phi(f') \geqq \phi(f)$; antitone, if $f' \geqq f$ implies $\phi(f') \leqq \phi(f)$; monotone, if ϕ is either isotone or antitone.

As is customary, when ϕ is on a domain F, we shall write $\phi(F)$ for the set $\{\phi(f) \mid f \text{ in } F\}$, that is the range of ϕ.

By a net of elements of a set S we shall mean a function whose domain is a directed set and whose range is contained in S. (The word "net" was introduced in this sense by J. L. Kelley, (Kelley [1])).

In discussing limit processes we frequently encounter statements $P(\alpha)$, concerning elements of a directed set A, and incorporated in clauses "there exists an α' in A such that $P(\alpha)$ is true for all elements α of A such that $\alpha \geqq \alpha'$ ". In such cases, this clause may be abbreviated to "$P(\alpha)$ is eventually true," or "eventually, $P(\alpha)$."

(3.1) DEFINITION. Let $(f(\alpha) \mid \alpha \text{ in } A)$ be a net of elements of a partially ordered set F. The net f is o-convergent if there exist subsets M, N of F such that

(a) M is directed by $\geqq$, and N is directed by $\leqq$;

(b) $\bigvee M$ and $\bigwedge N$ exist and are equal;

(c) for each m in M and each n in N it is eventually true that $m \leqq f(\alpha) \leqq n$.

In this case we define

$$\operatorname*{o-lim}_{\alpha \text{ in } A} f(\alpha) = \bigvee M ,$$

and say that $f(\alpha)$ is o-convergent to $\bigvee M$.

REMARK 1. If f is o-convergent, its o-limit is uniquely determined. For if M and N satisfy (3.1), and M' and N' also satisfy it, then for each m' in M' and each n in N it is eventually true that $m' \leqq f(\alpha)$ and that $f(\alpha) \leqq n$; hence $m' \leqq n$. This implies $\bigvee M' \leqq \bigwedge N = \bigvee M$. Likewise $\bigvee M \leqq \bigvee M'$, so the two limits are equal.

REMARK 2. If f is isotone and $\bigvee f(A)$ exists, f is o-convergent to $\bigvee f(A)$. For (3.1) holds if we take $M = f(A)$ and $N = \{\bigvee f(A)\}$. Likewise, if f is antitone and $\bigwedge f(A)$ exists, f is o-convergent to $\bigwedge f(A)$.

REMARK 3. If f is constant, it is o-convergent to its (unique) functional value. This follows from the preceding remark.

For some, but not all, nets we can define related upper and lower limits, as follows.

(3.2) DEFINITION. Let $(f(\alpha) : \alpha \text{ in } A)$ be a net of elements of a partially ordered set F. If for each α in A the supremum $n(\alpha) = \bigvee\{f(\alpha') : \alpha' \text{ in } A \text{ and } \alpha' \geqq \alpha\}$ exists, and also $\bigwedge\{n(\alpha) : \alpha \text{ in } A\}$ exists, we define $\operatorname*{o-lim\,sup}_{\alpha \text{ in } A} f(\alpha)$ to be $\bigwedge\{n(\alpha) : \alpha \text{ in } A\}$. The lower limit $\operatorname*{o-lim\,inf}_{\alpha \text{ in } A} f(\alpha)$ is defined dually.

To simplify the notation we shall usually abbreviate the symbol $\operatorname*{o-lim\,sup}_{\alpha \text{ in } A} f(\alpha)$ to $\text{o-lim sup}_\alpha f(\alpha)$ or to o-lim sup f, and likewise for the lower limit.

(3.3) COROLLARY. If $(f(\alpha):\alpha$ in $A)$ is a net of elements of a partially ordered set F, and o-lim $\sup_\alpha f(\alpha)$ and o-lim $\inf_\alpha f(\alpha)$ exist, then

(i) o-lim $\sup_\alpha f(\alpha) \geqq$ o-lim $\inf_\alpha f(\alpha)$;

and

(ii) o-$\lim_\alpha f(\alpha)$ exists if and only if o-lim $\sup_\alpha f(\alpha)$ and o-lim $\inf_\alpha f(\alpha)$ are equal, and in that case o-$\lim_\alpha f(\alpha)$ is equal to their common value.

Conclusion (i) is evident. Consider (ii). If the upper and lower limits are equal, the set $\{n(\alpha):\alpha$ in $A\}$ defined in (3.2) and the dually defined set $\{m(\alpha):\alpha$ in $A\}$ have all the properties specified in (3.1). Conversely, if o-$\lim_\alpha f(\alpha)$ exists, let M and N be as stated in (3.1). For each n in N, $f(\alpha)$ is eventually $\leqq n$, so for some α' we have $n(\alpha') \leqq n$, and therefore o-lim $\sup_\alpha f(\alpha) \leqq n$. Hence o-lim $\sup_\alpha f(\alpha) \leqq \bigwedge N =$ o-$\lim_\alpha f(\alpha)$. Similarly o-lim $\inf_\alpha f(\alpha) \geqq$ o-$\lim_\alpha f(\alpha)$, which with (i) completes the proof.

In particular, if F is a complete lattice, the upper and lower limits of every net of elements of F are both defined.

There is some interest in finding the relationship between convergence of a net in a product-space and convergence of its projections into the factor spaces.

(3.4) THEOREM. Let F_δ, δ in D be a collection of spaces, F_δ being partially ordered by $>_\delta$; and let $[F,>]$ be the cartesian product $\times_\delta [F_\delta, >_\delta]$. Let $f = (f(\alpha):\alpha$ in $A)$ be a net of elements of F. In order that f be o-convergent to an element f^* of F, it is necessary and sufficient that

(i) there exist elements h', h'' of F such that eventually, $h' \leqq f(\alpha) \leqq h''$, and

(ii) for each δ in D, the projection f_δ of f on F_δ shall be o-convergent to the projection f^*_δ of f^* on F_δ.

If D is a finite set, (i) may be omitted.

Suppose first that o-lim $f(\alpha) = f^*$. Let M,N be as in (3.1). We choose for h' any element of M and for h'' any element of N; (i) is satisfied. If M_δ, N_δ are the projections of M,N respectively on F_δ, they are directed by $\geqq_\delta$, $\leqq_\delta$ respectively, and by (2.6) $\bigvee M_\delta = \bigwedge N_\delta = f^*_\delta$. If m', n' are in M_δ, N_δ respectively, there exist m,n in M,N respectively whose projections on F_δ are m', n' respectively, so eventually

$m' \leqq f_\delta(\alpha) \leqq n'$. Hence (ii) holds.

Conversely, suppose (i) and (ii) satisfied. For each δ in D there are sets M_δ, N_δ directed by $\geqq_\delta, \leqq_\delta$ respectively, having $\vee M_\delta = \wedge N_\delta = f^*_\delta$, and for each m_δ in M_δ and n_δ in N_δ it is eventually true that $m_\delta \leqq_\delta f_\delta(\alpha) \leqq_\delta n_\delta$. For each finite subset D_0 of D we form all elements m of F such that m_δ is in M_δ for δ in D_0, $m_\delta = h'_\delta$ for δ in $D - D_0$. The aggregate of all such elements, for all choices of finite subsets of D, we call M. [If D is finite we can take D_0 to be D and omit all mention of h' and h"]. This is easily seen to be directed by $\geqq$ and to have f^* for supremum; and if m is in M, it is eventually true that $f(\alpha) \geqq m$. The set N can be defined analogously, so f is o-convergent to f^*.

Two special cases occur often enough in later pages to make it worth while to mention them as examples:

(3.5) COROLLARY. If H is the set of all extended-real-valued functions on a domain D_H, with the partial ordering that $h_1 \leqq h_2$ if $h_1(x) \leqq h_2(x)$ for all x in D_H, then a directed system $(h_\alpha : \alpha$ in A) of elements of H is convergent to an element h_0 of H if and only if $h_\alpha(x)$ converges to $h_0(x)$ for each x in D_H. If H is the set of all real-valued functions on a domain D_H, ordered as just described, then a directed system $(h_\alpha : \alpha$ in A) of elements of H is o-convergent to an element h_0 of H if and only if it is "eventually dominated" (that is, there exist elements h' and h" of H such that $h' \leqq h_\alpha \leqq h''$ for all α beyond a certain α') and $h_\alpha(x)$ converges to $h_0(x)$ for all x in D_H.

The second statement is a specialization of (3.4). So is the first, if we take h' constantly $-\infty$ and h" constantly $+\infty$.

Again, let F be a lattice, and let F^n be the partially-ordered product of n copies of F. Let $L = (L(a_1, \ldots, a_n)$: $a_1, \ldots, a_n$ in F) be a lattice combination defined on n-uples of elements of F. Then L maps F^n on F. It is easy to see that the mapping is isotone. We now prove

(3.6) THEOREM. With the notation above, if $(a_1(\alpha) : \alpha$ in A), ..., $(a_n(\alpha) : \alpha$ in A) are nets of elements with the respective o-limits $a'_1, \ldots, a'_n$, the net $(L(a_1(\alpha), \ldots, a_n(\alpha)) : \alpha$ in A) is o-convergent to $L(a'_1, \ldots, a'_n)$.

For each number j in the set $1, \ldots, n$, there are sets M_j, N_j directed by $\geqq$, respectively, having $\bigvee M_j = \bigwedge N_j = a'_j$ and such that for each m_j in M_j and each n_j in N_j it is eventually true that $m_j \leqq a_j(\alpha) \leqq n_j$. Define $M = X_j M_j$, $N = X_j N_j$. These are easily seen to have the properties needed to establish the conclusion.

It is not unusual to meet the situation in which some theorem can be proved about an o-convergent net provided that its domain A is countable. Accordingly we introduce a symbol to indicate this state of affairs, as follows.

(3.7) DEFINITION. A net $(f(\alpha):\alpha$ in $A)$ is σo-convergent to a limit f^* provided that

(i) it is o-convergent to f^* and

(ii) the domain A is countable.

As we see from (2.4), this differs but little from requiring the net to be an ordinary sequence. Nevertheless, there is a possibly useful gain in generality. For example, a double sequence $(f_{ij}: i,j = 1, 2, \ldots)$ of reals having a limit as i and j tend to ∞ is σo-convergent.

We adopt J. L. Kelley's definition (Kelley [1]) of a subnet.

(3.8) DEFINITION. Let $f = (f(\alpha):\alpha$ in $A)$ be a net. Let $(\alpha(\beta) \mid \beta$ in $B)$ be a net of elements of A such that for each α_0 in A there exists a β_0 in B such that $\alpha(\beta) \geqq \alpha_0$ whenever $\beta \geqq \beta_0$. Then the net $(f(\alpha(\beta)):\beta$ in $B)$ is a subnet of the net f.

With this definition it is clear that:

(3.9) If f is o-convergent to a limit f_0, so is every subnet of F.

Along with any type of convergence of sequences one can define (Kantorovitch [1]) a related type of convergence, called *-convergence by Kantorovitch, in the following manner. A sequence $f_1, f_2, \ldots$ contains a sub-subsequence which converges to f_0. In the present setting we can make some (but not much!) use of a related concept, for which we shall use the name o*-convergence, but we introduce two modifications. First, we consider nets instead of ordinary sequences. Second, for each subnet of f we shall ask less than the existence of a convergent subnet; we shall ask, roughly stated, that f_0 be a contact point of the aggregate of values of the subnet. Precisely, we make the following definition.

(3.10) DEFINITION. Let $f = (f(\alpha):\alpha$ in $A)$ be a net of elements of a partially ordered set F, and let f_o be an element of F. The net f is o*-convergent to f_o if whenever $(f(\alpha(\beta)):\beta$ in $B)$ is a subnet of f, there exists a net $(\beta(\gamma):\gamma$ in $\Gamma)$ of elements of B such that the net $(f(\alpha(\beta(\gamma))):\gamma$ in $\Gamma)$ is o-convergent to f_o.

Note that the last-mentioned net need not be a subnet of f, since we have not required that $\beta(\gamma)$ be eventually beyond each element of B. Also, it is clear that if f is o-convergent to f_o, it is also o*-convergent to f_o.

The concept of o*-convergent can be modified to furnish another (and to us rather more useful) type of convergence related to it as σo-convergence is related to o-convergence. All that is needed is to change the word "o-convergent" near the end of (3.9) to "σo-convergent." The result will be called σo*-convergence. Because of (2.4) we can change the form of the definition slightly without changing its content; the result is

(3.11) DEFINITION. Let $f = (f(\alpha):\alpha$ in $A)$ be a net of elements of a partially ordered set F, and let f_o be an element of F. The net f is σo*-convergent to f_o if whenever $(f(\alpha(\beta)):\beta$ in $B)$ is a subnet of F, there exists a sequence $\beta_1, \beta_2, \ldots$ of elements of B such that $\text{o-}\lim_{n\to\infty} f(\alpha(\beta_n)) = f_o$.

Clearly all σo*-convergent nets are o*-convergent. The following lemma establishes the converse in certain special cases.

(3.12) LEMMA. Let $f = (f(\alpha):\alpha$ in $A)$ be a net of elements of a partially ordered set F, and let f be o*-convergent to an element f_o of F. Then if either of the following hypotheses

(i) A contains a countable cofinal subset,

(ii) F consists of all real-valued or all extended-real-valued functions on a countable set D_f

is satisfied, the net f is σo*-convergent to f_o.

Let us first establish an auxiliary statement.

(A). If $(f(\alpha):\alpha$ in $A)$ is o-convergent to f_o, and its range is countable, there exists a sequence $\alpha_1, \alpha_2, \ldots$ of elements of A such that $\text{o-lim}_{n \to \infty} f(\alpha_n) = f_o$.

If for some c in the range of f the equation $f(\alpha) = c$ holds on a cofinal subset $A[c]$ of A, then $(f(\alpha):\alpha$ in $A[c])$ is a subnet of f on which f is constantly equal to c and is still o-convergent to f_o, so by the third remark after (3.1) $c = f_o$. If we set each α_n equal to an arbitrary member of $A[c]$ the conclusion holds. If for no c in the range of f does the equation $f(\alpha) = c$ hold on a cofinal subset, we first enumerate the elements $c_1, c_2 \ldots$ of the range of f. For each integer j there is an element α'_j of A such that $\alpha \leqq \alpha'_j$ for all α in the set $A[c_j]$ on which $f(\alpha) = c_j$. By induction, we choose a sequence of elements $\alpha_1, \alpha_2, \ldots$ of A such that for each j, $\alpha_j \geqq \alpha'_j$ and $\alpha_j \geqq \alpha_{j-1}$. These are cofinal in A; for each element α of A is in some $A[c_k]$, hence satisfies $\alpha \leqq \alpha'_k \leqq \alpha_k$. They are directed, in fact linearly ordered. So $(f(\alpha_j): j = 1, 2, \ldots)$ is a subnet of f, and is o-convergent to f_o. This establishes statement (A).

Now assume that (i) holds. Let $\alpha_1, \alpha_2, \ldots$ be cofinal in A. If $(f(\alpha(\beta)):\beta$ in $B)$ is a subnet of f, for each positive integer j there is an element $\beta(j)$ of B such that $\beta(j) \geqq \alpha_j$, $j = 1, 2, \ldots$. Then $(f(\alpha(\beta(j))): j = 1, 2, \ldots)$ is a subnet of f, and by the hypothesis of o*-convergence there is a net of integers $(j(\gamma):\gamma$ in $\Gamma)$ such that $(f(\alpha(\beta(j(\gamma)))):\gamma$ in $\Gamma)$ is o-convergent to f_o. But the range of this net is countable, since j takes on only integral values; so by statement (A) there is a sequence $\gamma_1, \gamma_2, \ldots$ of elements of Γ such that $\text{o-lim}_{N \to \infty} f(\alpha(\beta(j(\gamma_n)))) = f_o$. Thus if we define $\beta_n = \beta(j(\gamma_n))$ the requirements of definition (3.11) are satisfied.

If (ii) holds, let $(f(\alpha(\beta)):\beta$ in $B)$ be a subnet of f. By definition, there exists a net $(\beta(\gamma):\gamma$ in $\Gamma)$ of elements of B such that $f(\alpha(\beta(\gamma)))$ is o-convergent to f_o. As noted in (3.5), there exist elements h', h'' of F such that the inequalities $h' \leqq f(\alpha(\beta(\gamma))) \leqq h''$ hold for all γ beyond a certain γ_o of Γ; the other elements of Γ we discard. Also, at each of the points $x_1, x_2, \ldots$ D_f the value of $f(\alpha(\beta(\gamma)))$ tends to the value of f_o at that point. Now for each positive integer n we choose a γ_n in Γ such that at each point x_j of D_f with $j \leqq n$ the value of $f(\alpha(\beta(\gamma_n)))$ is in the $(1/n)$-neighborhood of the value of f_o: and we define $\beta_n = \beta(\gamma_n)$. Then $f(\alpha(\beta_n))$ converges pointwise to f_o, being also comprised between h' and h''. Hence $\text{o-lim}_{n \to \infty} f(\alpha(\beta_n)) = f_o$, completing the proof.

It is now quite easy to show the following relationships between the four types of convergence.

(3.13) **THEOREM.** Let f be a net of elements of a partially ordered set F, and let f_0 be an element of F. If f is σo-convergent to f_0, it converges to f_0 in all four modes of convergence. If f converges to f_0 in any one of the four modes, it is o*-convergent to f_0.

If f is σo-convergent to f_0, its domain is countable, so by (3.10) it is σo*-convergent to f_0. The other statements are obvious.

§ 4. CLOSURE

With each type of convergence one naturally associates a type of closure. For example, if S is a subset of a partially ordered set F, we shall say that S is "closed under o-convergence" if it is true that whenever f is a net of elements of S which is o-convergent to an element f_0 of F, f_0 belongs to S. The analogue holds if we replace o-convergence by σo-convergence, o*-convergence or σo*-convergence.

In addition to these types of closure, there is another type that appears useful in studying partially ordered sets; this is closely related to Dedekind completeness, and we shall accordingly call it Dedekind closure.

(4.1) **DEFINITION.** Let S be a subset of a partially ordered set F. S is Dedekind [σ-]closed if for every [countable] subset K of S directed by $\geqq$ and having a supremum in F, $\bigvee K$ is in S, and for every [countable] subset K of S directed by $\leqq$ and having an infimum in F, $\bigwedge K$ is in S.

In the next theorem we collect some fairly trivial relations between the various types of closure.

(4.2) **THEOREM.** Let S be a subset of a partially ordered set F.

(i) S is closed under o-convergence if and only if it is closed under o*-convergence.

(ii) S is closed under σo-convergence if and only if it is closed under σo*-convergence.

(iii) If S is closed under o-convergence, it is closed under σo-convergence.

(iv) If S is Dedekind closed, it is Dedekind σ-closed.

(v) If S is closed under o-convergence, it is Dedekind closed.

(vi) If S is closed under σo-convergence, it is Dedekind σ-closed.

An element of F which is the limit of an o-convergent net of elements of S is the limit of an o*-convergent net, namely the same net; and an element which is the limit of an o*-convergent net of elements of S is also by (3.9) the limit of an o-convergent net of elements of S. So (i) clearly holds. The proof of (ii) is similar; and (iii) is obvious, and likewise (iv). If S is closed under o-convergence and K is a subset of S directed by $\geqq$ and having a supremum k' in F, then $\phi = (\phi(k) = k:k \text{ in } K)$ is an isotone net which o-converges to k', by the second remark after (3.1). So k' is in K. The analogue also holds for subsets directed by $\leqq$, so S is Dedekind closed, and (v) is established. The proof of (vi) is essentially the same.

To show that the four types of closure mentioned in (v) and (vi) are distinct, let F be the set of extended-real-valued functions on the real number system. The functions of Baire form a set which is Dedekind σ-closed and also closed under σo-convergence, but is neither Dedekind closed nor closed under o-convergence. The set S of functions defined by $f_n(x) = [\sin nx]/n$, $n = k, 2, \ldots,$ is Dedekind closed and Dedekind σ-closed, because no subset of S directed by $\geqq$ or by $\leqq$ can have more than one element in it; but since the zero-function is not in S, S is not closed under o-convergence or under σ o-convergence. However, there is a non-trivial case in which Dedekind σ-closure coincides with closure under σo-convergence, and likewise with the σ's omitted.

(4.3) THEOREM. Let F be a Dedekind [σ-]complete partially ordered set, and let S be a lattice embedded in F having an upper bound and a lower bound in F. Then the following statements are equivalent.

(i) S is Dedekind [σ-]closed.

(ii) S is closed under [σ]o-convergence.

(iii) S is a [σ-]complete lattice.

Assume that (i) holds. Let K be a [countable] subset of S, and let K* be the set of infima of finite subsets of K. As in proving (2.3), K* is directed by $\leqq$; and since S is a lattice, K* is contained in S. By (i), $\bigwedge K^*$ is in S, so by (1.2) $\bigwedge K$ is in S. Similarly $\bigvee K$ is in S, and (iii) holds. Next assume that (iii) holds and that $f = (f(\alpha):\alpha$ in $A)$ is a net of elements of S which is $[\sigma]$o-convergent to an element f_0 of F. By (iii), for each α' in A the element $\bigwedge\{f(\alpha):\alpha$ in A and $\alpha \geqq \alpha'\}$ is in S, and so is the supremum of these elements, which is f_0 by (3.3). So (ii) holds. If (ii) holds so does (i), by (4.2), and the proof is complete.

§ 5. CONTINUITY

If ϕ is a function on a subset S of a partially ordered set F, and the values of ϕ lie in a partially ordered set G, we could use any one of our four modes of convergence (o, σo, o* and σo*) in each space, and thus define sixteen kinds of continuity. But we rarely have occasion to refer to types of continuity in which different modes of convergence are used in F and G, so to avoid complexity we introduce names only for the four kinds of continuity in which the same mode of convergence is used in F and in G:

(5.1) DEFINITION. If the domain D of ϕ is contained in a partially ordered set F and its range is in a partially ordered set G, then ϕ is $[\sigma]$o-continuous at a point s' of D if for each net $(s(\alpha):\alpha$ in $A)$ of elements of D which is $[\sigma]$o-convergent to s', the net $(\phi(s(\alpha)):\alpha$ in $A)$ is $[\sigma]$o-convergent to $\phi(s')$ in G. Also ϕ is $[\sigma]$o-continuous in D if it is $[\sigma]$o-continuous at each point s' of D. The definition of o*-continuity and σo*-continuity are analogous, $[\sigma]$o-convergence being replaced by $[\sigma]$o*-convergence in each place where it occurs.

There are some simple relations among the four kinds of continuity.

(5.2) THEOREM. If ϕ is o-continuous at a point s' of D, it is also σo-continuous, o*-continuous and σo*-continuous at s'. If it is σo-continuous at s' it is also σo*-continuous at s'.

If ϕ is o-continuous and $(s(\alpha):\alpha$ in $A)$ is a net of elements of D and is σo-convergent to an element s' of D, it is also o-convergent, so $(\phi(s(\alpha)):\alpha$ in $A)$ is o-convergent to $\phi(s')$. Since A is countable, it is also σo-convergent. If $(s(\alpha):\alpha$ in $A)$ is a net of elements of D and is o*-convergent to s' in D, let $(\phi(s(\alpha(\beta))):$ β in $B)$ be a subnet of $(\phi(s(\alpha)):\alpha$ in $A)$. Then $(s(\alpha(\beta)):$ β in $B)$ is a subnet of $(s(\alpha):\alpha$ in $A)$, so there is a net

$(s(\alpha(\beta(\gamma))):\gamma$ in $\Gamma)$ which is o-convergent to s'. Then $(\phi(s(\alpha(\beta(\gamma)))):\gamma$ in $\Gamma)$ is o-convergent to $\phi(s')$, so ϕ is o*-continuous.

If the net $(s(\alpha):\alpha$ in $A)$ is σo*-convergent to s', the set Γ can be chosen countable, so the σo-continuity of ϕ implies that $(\phi(s(\alpha(\beta(\gamma)))):\gamma$ in $\Gamma)$ is σo-convergent to $\phi(s')$, and ϕ is σo*-convergent.

From (3.6) we have at once the useful result:

(5.3) THEOREM. If F is a lattice and L a lattice combination defined for n-uples of elements of F, L is an o-continuous mapping of F^n into F.

Our aim is to extend the domain of an isotone function to a larger function with useful closure properties. An o-continuous function ϕ cannot always be thus extended. For example, if $\phi(x) = 1$ for all rational $x > \sqrt{2}$ and $\phi(x) = 0$ for all rational $x < \sqrt{2}$, we see that ϕ is continuous on the rationals, but it has no continuous extension to any domain containing $\sqrt{2}$. We wish to restrict our attention to functions without "jumps." Accordingly, we introduce the following definition.

(5.4) DEFINITION. Let ϕ be a function whose domain D is contained in a partially ordered set F and whose range is contained in a partially ordered set G. The function ϕ is [σ]-smoothly rising if:

(i) it is isotone, and

(ii) if S_1 and S_2 are [countable] subsets of D directed by $\geqq$ and $\leqq$ respectively, and $\bigvee S_1$, $\bigvee \phi(S_1)$, $\bigwedge S_2$ and $\bigwedge \phi(S_2)$ exist, and $\bigvee S_1 \geqq \bigwedge S_2$, then $\bigvee \phi(S_1) \geqq \bigwedge \phi(S_2)$.

The definition of a [σ]-smoothly descending function is obtained from this by the obvious reversals of order.

The following corollary is practically self-evident.

(5.5) COROLLARY. Let ϕ be an isotone function on a Dedekind [σ]closed domain D. Assume that whenever S_1 is a [countable] subset of D directed by $\geqq$, and $\bigvee S_1$ and $\bigvee \phi(S_1)$ exist, then $\phi(\bigvee S_1) = \bigvee \phi(S_1)$, and dually for sets directed by $\leqq$. Then ϕ is [σ]smoothly rising on D.

(5.6) COROLLARY. Let ϕ be a $[\sigma\text{-}]$smoothly rising function whose domain D is a $[\sigma\text{-}]$complete lattice in a partially ordered set F, and whose range is in a Dedekind $[\sigma\text{-}]$complete partially ordered set G. Then ϕ is $[\sigma]$o-continuous.

Let $(s(\alpha):\alpha$ in $A)$ be a net of elements of D which is $[\sigma]$o-convergent to an element s_o of D. Let N be the set $\{n(\alpha):\alpha$ in $A\}$ defined in (3.2), and let M be the set $\{m(\alpha):\alpha$ in $A\}$ dually defined. [These sets are countable.] As in proving (3.3), M and N have the properties specified in the definition (3.1), and $\bigvee M = s_o = \bigwedge N$. Since ϕ is isotone, the set $\{\phi(m(\alpha)):\alpha$ in $A\}$ is directed by $\geqq$ and has $\phi(s_o)$ as upper bound, so $\bigvee\phi(M) \leqq \phi(s_o)$. Likewise $\bigwedge\phi(N) \geqq \phi(s_o)$. But the set consisting of s_o alone is directed by $\leqq$ and by $\geqq$, and since ϕ is $[\sigma\text{-}]$smoothly rising we have $\bigvee\phi(M) \geqq \phi(s_o) \geqq \bigwedge\phi(N)$. Hence equality holds. For each $m(\alpha')$ in M and each $n(\alpha'')$ in N we eventually have $m(\alpha') \leqq \phi(s(\alpha)) \leqq n(\alpha'')$, so $[\sigma]\text{o-lim}\ \phi(s(\alpha)) = \phi(s_o)$.

(5.7) COROLLARY. If ϕ is a smoothly rising function whose domain is all of a partially ordered set F and whose range is contained in a Dedekind complete partially ordered set G, then ϕ is o-continuous.

Let $(s(\alpha):\alpha$ in $A)$ be a net o-convergent to an element s_o of F, and let M and N be subsets of F having the properties specified in the definition (3.1). Since ϕ is isotone, $\phi(M)$ and $\phi(N)$ are directed by $\geqq$ and $\leqq$ respectively. Also $\phi(s_o)$ is an upper bound for $\phi(M)$ and a lower bound for $\phi(N)$. Hence $\bigvee\phi(N)$ and $\bigwedge\phi(N)$ exist and satisfy $\bigvee\phi(M) \leqq \phi(s_o) \leqq \bigwedge\phi(N)$. Because ϕ is smoothly rising, the reverse inequalities also hold, which establishes equality. For each m in M and n in N, $s(\alpha)$ is eventually between m and n, so $\phi(s(\alpha))$ is eventually between $\phi(m)$ and $\phi(n)$. Hence by definition (3.1) $\text{o-lim}\ \phi(s(\alpha)) = \phi(s_o)$.

Among the isotone functions on partially ordered sets, an especially important subclass consists of those smoothly rising functions whose range is in the real number system. For these we introduce a symbol.

(5.8) DEFINITION. If F is partially ordered, R_F is the class of all smoothly rising (hence o-continuous and isotone) real-valued functions whose domain is all of F.

In topology, some especially desirable properties are possessed by the "normal" spaces, which in essence are those in which there are enough continuous real-valued functions to determine the topology. The analogues in the present situation would be those partially-ordered sets in which there are enough smoothly-rising real valued functions to determine the order. For such sets we borrow the name "normal" from their topological analogues, as follows.

(5.9) DEFINITION. A partially ordered set F is normal if for each pair f_1, f_2 of elements of F such that the inequality $r(f_1) \leqq r(f_2)$ holds for each r in the class R_F, it is true that $f_1 \leqq f_2$.

We shall make use of this concept later, in Section 14 and thereafter.

§ 6. PARTIALLY ORDERED GROUPS

The partially ordered sets most frequently useful to us are those which have one or more operations of an algebraic nature defined on them. The simplest such systems are those with one operation.

Here the most obvious structures to study would be partially ordered groups. However, the need for a little more latitude is shown by a simple example. The real numbers are merely a conditionally complete lattice; we convert them into a complete lattice by adjoining $-\infty$ and ∞. But this extended real number system is no longer a group under addition. If we wish the advantages of both, we must study the group of real numbers as a part of the complete lattice of extended reals. We thus seek to connect the group operation with the order-relation in such a way that bound-relations are preserved under addition and reversed under subtraction, and we must do this without assuming the existence of suprema and infima and without assuming that addition and subtraction are defined in the whole partially ordered set. The result is the definition below; in (6.6) we show its simplifications when infima and suprema exist.

(6.1) DEFINITION. A group with a set F_+ of elements and a (not necessarily commutative) operation $+$ is a group embedded in a partially ordered set F if

(a) F_+ is contained in F;

(b) if a, b, x, y and z are in F_+, and each lower bound (in F) for the pair $\{y,z\}$ is $\leqq x$, then each upper bound (in F) for the pair $\{a - y + b, a - z + b\}$ is $\geqq a - x + b$, and dually; and

(c) the function $(x + y: x \text{ in } F_+, y \text{ in } F_+)$ is smoothly rising on $F_+ \times F_+$, and the function $(-x: x \text{ in } F_+)$ is smoothly descending on F_+.

The identity, or zero element, of F_+ will be called θ or θ_{F_+}.

If F_+ coincides with F, we say that F is a partially ordered group under $+$.

From this definition we deduce some easy but useful consequences. In all these corollaries we assume without further mention that (6.1) is satisfied; but it is worth noticing that we make no use of (6.1c) until Theorem (6.7).

(6.2) COROLLARY. Let a, b, x and y be members of F_+.

(i) If $x \geqq y$, then $a + x + b \geqq a + y + b$ and $a - x + b \leqq a - y + b$.

(ii) If $x \geqq a$ and $y \geqq b$, then $x + y \geqq a + b$.

(iii) The four statements $x \geqq y$, $x - y \geqq \theta$, $-y + x \geqq \theta$, $-y \geqq -x$ are equivalent.

If $x \geqq y$, by setting $z = y$ in (6.1b) we obtain $a - x + b \leqq a - y + b$. With $a = b = \theta$, this implies $-x \leqq -y$. Now $x \geqq y$ implies $-x \leqq -y$; replacing x by $-y$ and y by $-x$ shows that this is an equivalence. By the part of (i) already established, $-b-x-a \leqq -b-y-a$, which by the equivalence completes the proof of (i). By two applications of (i), $x + b \geqq a + b$ and $x + y \geqq x + b$, whence (ii) follows; and (iii) is an obvious consequence of (ii).

(6.3) COROLLARY. Let a, b, x, y be members of F_+. If the pair $\{x,y\}$ has either a supremum or an infimum in F which is a member of F_+, then all four pairs $\{x,y,\}$ $\{-x,-y\}$, $\{a + x + b, a + y + b,\}$ $\{a - x + b, a - y + b\}$ have both infima and suprema, all belonging to F_+, and the equations

$$a + (x \vee y) + b = (a + x + b) \vee (a + y + b)$$

$$a - (x \vee y) + b = (a - x + b) \wedge (a - y + b)$$

$$(-x) \vee (-y) = -(x \wedge y)$$

and their duals are satisfied. In particular, if F_+ is a commutative group, $x \vee y + x \wedge y = x + y$.

Assume first that $x \vee y$ exists and is in F_+. By (6.2111), $-(x \vee y)$ is a lower bound for $-x$ and $-y$, so by (6.21) $a-(x \vee y) + b$ is a lower bound for $a - x + b$ and $a - y + b$. Every upper bound for $\{x,y\}$ is $\geqq x \vee y$, so by (6.1b) every lower bound for $\{a - x + b, a - y + b\}$ is $\leqq a - (x \vee y) + b$. That is, $a - (x \vee y) + b = (a - x + b) \wedge (a - y + b)$. In particular, if we set $a = y$ and $b = x$, we find that $x \wedge y$ exists and is $y - (x \vee y) + x$, being therefore in F_+. When addition is commutative this yields the last equation in the conclusion. Since $x \wedge y$ exists, we can dualize the proof above and show that $(a - x + b) \vee (a - y + b)$ exists and is $a - (x \wedge y) - b$. Letting $a = b = \theta$ gives us the second-last equation in the conclusion. This permits applying the above proof to $-x$ and $-y$; replacing x by $-x$ and y by $-y$ in the equation just established furnishes the remaining conclusion.

The equivalents of f^+, f^- and $|f|$ as usually understood in function-spaces or in lattices do not always exist in partially ordered groups, but we follow the usual path when this is possible.

(6.4) DEFINITION. Let F_+ be a group embedded in a partially ordered space F. If f is a member of F_+ such that $f \vee \theta$ and $(-f) \vee \theta$ both exist and are in F_+, we define

$$f^+ = f \vee \theta,$$

$$f^- = (-f) \vee \theta,$$

$$|f| = f^+ + f^-.$$

REMARK. By (6.3), if either of the suprema $f \vee \theta$ and $(-f \vee \theta)$ exists and belongs to F_+, so does the other.

(6.5) COROLLARY. When f^+ and f^- are defined, the relations $f^+ \geqq \theta$, $f^- \geqq \theta$, $|f| \geqq \theta$, $f^+ - f^- = -f^- + f^+ = f$ are satisfied.

The inequalities are obvious. By hypothesis, $(-f) \vee \theta$ exists, so by (6.3) $f + (-f) \vee \theta + \theta = (f + [-f] + \theta) \vee (f + \theta + \theta)$, or $f + f^- = f^+$. Also by (6.3), $\theta + (-f) \vee \theta + f = (\theta + [-f] + f) \vee (\theta + \theta + f)$, or $f^- + f = f^+$. These imply the conclusion.

(6.6) THEOREM. Let F_+ be a group under $+$ and be contained in a partially ordered set F. If each pair of elements of F_+ has a supremum and an infimum in F (not necessarily in F_+), (6.1b) is equivalent to the statement

(6.1b') If a,b,x,y,z are in F_+ and $x \geqq y \wedge z$, then $(a - x + b) \leqq (a - y + b) \vee (a - z + b)$, and dually.

When F_+ is a lattice embedded in F, (6.1b) is equivalent to

(6.1b") If a,b,x,y are in F_+ and $x \geqq y$, then $a - x + b \leqq a - y + b$.

Clearly (6.1b) implies (6.1b'), and it implies (6.1b") by (6.2). If $y \wedge z$ exists, the hypothesis "$s \leqq y$ and $s \leqq z$ implies $s \leqq x$" implies (with $s = y \wedge z$) that $x \geqq y \wedge z$, so by (6.1b') half the conclusion of (6.1b) holds; the other half is established dually. If F_+ is a lattice and (6.1b") holds, for each c,d,x,y in F_+ we have $c - (x \wedge y) + d \geqq c - x + d$ and $c - (x \wedge y) + d \geqq c - y + d$, so $c - (x \wedge y) + d \geqq (c - x + d) \vee (c - y + d)$. In the dual statement, which is similarly established, we first replace x by $a - x + b$ and y by $a - y + b$, then replace c by b and d by a; we obtain $b - (a - x + b) \vee (a - y + b) + a \leqq x \wedge y$, or $(a - x + b) \vee (a - y + b) \geqq a - (x \wedge y) + b$. The reversed inequality has already been established, so equality holds. Then the statement in (6.1b') concerning infima is satisfied. The other statement is established dually, so (6.1b') holds, and therefore so does (6.1b).

(6.7) LEMMA. If F_+ is a group embedded in a partially ordered set F, it has the property

(i) each pair f_1,f_2 of elements of F_+ which has an upper bound in F_+ also has a lower bound in F_+, and vice versa.

Moreover, every set F_+ with property (i), whether or not it is a group, has the further property

(ii) if $f_1, \ldots, f_n$ are elements of F_+, and each pair $\{f_{i-1},f_i\}$ $(i = 2, 3, \ldots, n)$ has an upper bound or a lower bound in F_+, then the whole set $\{f_1, \ldots, f_n\}$ has an upper bound and a lower bound in F_+.

Suppose to be specific that f_1 and f_2 have an upper bound b' in F_+. Then $b' - f_1 \geqq \theta$ and $-f_2 + b' \geqq \theta$, so $f_1 - b' \leqq \theta$ and $-b' + f_2 \leqq \theta$. If we define $b = f_1 - b' + f_2$, we have $b \leqq f_1$ and $b \leqq f_2$, and (i) is established. To establish (ii) use induction. By (i), (ii) holds for $n = 2$. Suppose it valid for $n = k - 1$, and let $f_1, \ldots, f_k$ be elements such that each pair of consecutive members has an upper or a lower bound (hence, by (i), has both) in F_+. By the induction hypothesis, the set $\{f_1, \ldots f_{k-1}\}$ has a lower bound c and an upper bound c' in F_+. Also by hypothesis, the pair $\{f_{k-1}, f_k\}$ has a lower bound d and an upper bound d' in F_+. Then f_{k-1} is an upper bound for the pair $\{c,d\}$, so by (i) the pair has a lower bound b in F_+. Clearly $b \leqq c \leqq f_i$ $(i = 1, \ldots, k - 1)$ and $b \leqq d \leqq f_k$, so b is a lower bound for $\{f_1, \ldots, f_k\}$. The existence of an upper bound is established dually, and by induction (ii) is valid.

The connection between bounds and the group operation is interesting enough to justify a digression, even though the results of the next few paragraphs are not essential to later proofs.

For each a in F_+ let F_a be the set of all b in F_+ such that a and b have a common upper or lower bound in F_+. If F_a and F_b have a common element c, and a',b' are arbitrary members of F_a, F_b respectively, by (6.7) the five elements a',a,c,b,b' have a common upper bound in F_+, so a' is in F_b and b' in F_a. Hence $F_b = F_a$. Thus the partially ordered group F_+ is classified into mutually exclusive classes F_a. If a is in the class F_θ containing θ, the pair $\{\theta,a\}$ has an upper bound b' and a lower bound b in F_+. Then $a = -[b' - a] + [b'] = [a - b] - [-b]$, the quantities in square brackets being $\geqq \theta$. So every element in F_θ can be represented in either of the forms $c - d$, $-d + c$, where c and d are $\geqq \theta$. Conversely, if x has either of these forms, c is an upper bound for x and θ, so x is in F_θ.

If a and b are in F_θ, there are elements c,d in F_+ such that $c \geqq a$, $c \geqq \theta$, $d \leqq \theta$, $d \leqq b$. Then $c - d \geqq \theta$ and $c - d \geqq a - b$, so $a - b$ is in F_θ. It follows that F_θ is a subgroup of F_+. It is in fact an invariant subgroup. For if a is in F_θ and x is in F_+, there exists an element b in F_+ such that $b \geqq \theta$ and $b \geqq a$. Then $-x + a + x \leqq -x + b + x$ and $-x + b + x \geqq -x + \theta + x = \theta$, so $-x + a + x$ is again in F_θ. For each set F_a, if b and c are both in F_a they have a common upper bound d in F_+, $d - b$ and $d - c$ are $\geqq \theta$, and so $b - c = -(d - b) + (d - c)$ is in F_θ. Conversely, if $b - a$ is in F_θ, $b - a$ and θ have a common upper bound c in F_+, so $b \leqq c + a$ and $a \leqq c + a$, and therefore b is in F_a. This shows that the sets F_a are the cosets into which F_+ is classified by the invariant subgroup F_θ.

The next theorem shows that (6.1c) follows from (6.1a,b), provided that F_+ has a certain closure property.

(6.8) **THEOREM**. Let F_+ be a group contained in a Dedekind complete partially ordered set F and satisfying (6.1a) and (6.1b). Assume that whenever a and b are in F_+ and $a \leqq b$, the set $\{x : x$ in F_+ and $a \leqq x \leqq b\}$ is Dedekind closed. Then (6.1c) holds, so that F_+ is a group embedded in F. Moreover, if S_1 and S_2 are subsets of F_+ directed by $\geqq$ and having upper bounds in F_+, and S is the set $\{x_1 + x_2 : x_1$ in S_1 and x_2 in $S_2\}$ and S_- is the set $\{-x : x$ in $S_1\}$, then $\vee S_1, \vee S_2, \vee S$ and $\wedge S_-$ all exist, and

$$\vee S = \vee S_1 + \vee S_2 ,$$

$$\wedge S_- = - \vee S_1 .$$

It is clear that (6.1c) follows from the statements in the last sentence, so this is all we need to prove.

The sets S_1, S_2 and S are directed by $\geqq$ and S_- is directed by $\leqq$. Also, if b_1 and b_2 are upper bounds in F_+ for S_1 and S_2 respectively, $b_1 + b_2$ is an upper bound for S and $-b_1$ is a lower bound for S_-. By the hypothesis of Dedekind completeness of F and Dedekind closure, $\vee S_1$, $\vee S_2$, $\vee S$ and $\wedge S_-$ all exist in F and belong to F_+. Clearly $\vee S_1 + \vee S_2$ is an upper bound for S, so $\vee S \leqq \vee S_1 + \vee S_2$. Let a_1 belong to S_1; for all a_2 in S_2 we have $\vee S \geqq a_1 + a_2$, whence $-a_1 + \vee S \geqq a_2$, and $-a_1 + \vee S \geqq \vee S_2$. Then $\vee S \geqq a_1 + \vee S_2$, so $\vee S - \vee S_2 \geqq a_1$. This holds for all a_1 in S_1, so $\vee S - \vee S_2 \geqq \vee S_1$, and $\vee S \geqq \vee S_1 + \vee S_2$. The reverse inequality was previously established, so equality holds, and the first equation established. The second is proved analogously.

The equations in the last conclusion of (6.8) can be established without assuming the Dedekind closure of the sets $\{x : x$ in F_+ and $a \leqq x \leqq b\}$ provided the sets S_1 and S_2 are finite, for then $\vee S_1$ is merely the greatest element in S_1. This and (6.8) might lead to the suspicion that (6.1c) is always a consequence of (6.1a,b). This is not so; if we omit the hypothesis concerning Dedekind closure in (6.8) the equations in the conclusion may fail even when all the infima and suprema mentioned in them exist. For example, let F_+ consist of the group of reals under addition, and let F consist of the reals with the usual order plus two elements q_{-1}, q_2 with the properties $q_{-1} > -1$, $q_{-1} < x$ whenever

$x > -1$; $q_2 < 2$, $q_2 > y$ whenever $y < 2$. Let S_1 consist of all $x < 1$, and let S_2 consist of 1 alone. Then $\bigvee S_1 = \bigvee S_2 = 1$, $\bigvee S = q_2 < 1 + 1$, $\bigwedge S_- = q_1 > -1$.

(6.9) COROLLARY. If F is a Dedekind-complete partially ordered group, the operation $+$ is o-continuous on $F \times F$ to F, and the function $(-x : x$ in $F)$ is also o-continuous.

This follows at once from (5.7).

§ 7. LINEAR SYSTEMS, MULTIPLICATIVE SYSTEMS AND ALGEBRAS WITHIN PARTIALLY ORDERED SETS

As usual, by introducing an operation of "scalar multiplication" in a commutative group we produce a "linear system." We shall always use the real numbers for our scalars.

(7.1) DEFINITION. A system consisting of a set F of objects, a binary operation $+$ and an operation of scalar multiplication, is a linear system if $(F,+)$ is a commutative group, and also:

(i) To each real α and each a in F, there corresponds a unique scalar product, denoted by αa or $a\alpha$; and, for all real numbers α, β and all a, b in F,

(ii) $(\alpha + \beta)a = \alpha a + \beta a$,

(iii) $\alpha(a + b) = \alpha a + \alpha b$,

(iv) $\alpha(\beta a) = (\alpha\beta)a$,

(v) $1a = a$.

It follows readily that $(-1)a = -a$ and $0a = \theta$ for all a in F.

To define a linear system embedded in a partially ordered set, it remains to relate the operations to the order-relation.

(7.2) DEFINITION. A linear system F_+ is a linear system embedded in a partially ordered set F if

(a) $(F_+,+)$ is a group embedded in F, and

(b) if a is in F_+, and α is real, and $\alpha \geqq 0$ and $a \geqq \theta$, then $a\alpha \geqq \theta$.

If F_+ coincides with F, we say that F is

a partially ordered linear system.

The following theorem is an obvious corollary.

(7.3) THEOREM. If F is a partially ordered linear system, and S is a subset of F having a supremum, and α is a positive number, then the set $\{\alpha a: a \text{ in } S\}$ has supremum $\alpha(\bigvee S)$, and the set $\{-\alpha a: a \text{ in } S\}$ has infimum $-\alpha(\bigvee S)$.

(7.4) THEOREM. Let F be a Dedekind σ-complete partially ordered linear system, and let a be a member of F. Then the following three statements are equivalent.

(i) $\operatorname{o-lim}_{n \to \infty} n^{-1}a$ exists,

(ii) $\operatorname{o-lim}_{n \to \infty} n^{-1}a$ exists and is equal to θ,

(iii) a and θ have a common upper or lower bound.

If (i) holds, by (3.1), (3.9) and (7.3) we have

$$\begin{aligned} \operatorname{o-lim} n^{-1}a &= \operatorname{o-lim} [(2n)^{-1}a] \\ &= \operatorname{o-lim} n^{-1}(a/2) \\ &= 2^{-1}[\operatorname{o-lim} n^{-1}a] , \end{aligned}$$

so (ii) holds. If (ii) holds, let M, N be as in (3.1). Let x be in N, and let n be a positive integer such that $n^{-1}a \leqq x$. Then $a \leqq nx$ and $\theta \leqq nx$, establishing (iii). Now assume that (iii) holds. By (6.7), there exist elements x, y such that $x \leqq \theta \leqq y$ and $x \leqq a \leqq y$. Let M be the set $\{n^{-1}x: n = 1, 2, \dots\}$ and N the set $\{n^{-1}y: n = 1, 2, \dots\}$. Then for each element $n^{-1}y$ of N we have, for all $m > n$, $m^{-1}a \leqq m^{-1}y \leqq n^{-1}y$, and likewise for elements of M. Since θ is a lower bound for N and F is Dedekind σ-complete, N has an infimum. As in the first sentence of this proof, $\bigwedge N = \theta$. Similarly $\bigvee M = \theta$, and (i) and (ii) are established.

If a binary multiplication with suitable properties is defined for all pairs of elements of the group, it becomes a ring. However, in order to cover some applications we need a less stringent requirement; we ask only that the binary multiplication (which as usual we indicate by juxtaposing the factors or by placing a dot between them) be defined for some pairs of elements. Apart from requirements concerning order, we place the following requirements on the operation.

(7.5) DEFINITION. An operation on pairs of elements of a commutative group $(F,+)$ is a binary multiplication in the group, or more briefly a multiplication, if the domain of the operation is a non-empty set contained in $F \times F$, and

(a) if a,b,c are in F and ab and ac are defined, so is $a(b-c)$, and $a(b-c) = ab - ac$;

(b) if a,b,c are in F and ba and ca are defined, so is $(b-c)a$, and $(b-c)a = ba - ca$.

It follows readily that if ab is defined, so are $a\theta$ and θb, and both are θ; and so are $a(-b)$ and $(-a)b$, and both are $-(ab)$. Also, if ab and ac are defined, so is $a(b + c)$, and it is equal to $ab + ac$; and likewise under change of order of factors.

For our purposes we must connect this multiplication with partial ordering.

(7.6) DEFINITION. A system consisting of a commutative group $[F_+,+]$ and a binary operation $\cdot$ is a group with multiplication embedded in a partially ordered set F if

(a) F_+ is a commutative group under $+$ embedded in the set F.

(b) The operation $\cdot$ is defined on a subset of the pairs of elements of F_+, and is a binary multiplication.

(c) If (a,b) is in the domain of the multiplication, and $a \geqq \theta$ and $b \geqq \theta$, then $ab \geqq \theta$.

(d) On the part of the domain of the binary multiplication consisting of those pairs (a,b) with $a \geqq \theta$ and $b \geqq \theta$, the binary multiplication is smoothly rising.

Moreover,

(i) if F_+ coincides with F, we say that F is a partially ordered group with a binary multiplication;

(ii) if the domain of the binary multiplication is all of $F_+ \times F_+$, we say that F_+ is a ring embedded in F; and

(iii) if F_+ coincides with F and the domain of the binary multiplication is all of $F \times F$, we say that F is a partially ordered ring.

Finally, we combine all three types of operations in a single system.

(7.7) DEFINITION. A linear system is a linear system with binary multiplication if there is an operation $\cdot$ such that $[F,+]$ and the operation $\cdot$ form a commutative group with multiplication, and moreover whenever α and β are real numbers and a and b are elements of F for which $a \cdot b$ is defined, $(\alpha a) \cdot (\beta b)$ is also defined, and is equal to $(\alpha\beta)(a \cdot b)$.

The combination with partial ordering suggests itself.

(7.8) DEFINITION. A linear system F_+ with (binary) multiplication is a linear system with multiplication embedded in a partially ordered set F if its elements with $+$ and scalar multiplication are a linear system embedded in F, and its elements with $+$ and $\cdot$ are a group with multiplication embedded in F. Moreover,

(i) if the elements of the system are all the elements of F, we say that F is a partially ordered linear system with multiplication;

(ii) if the domain of the binary multiplication is all of $F_+ \times F_+$, we say that F_+ is an algebra embedded in F; and

(iii) if F_+ is all of F and the domain of the binary multiplication is all of $F \times F$, we say that F is a partially ordered algebra.

CHAPTER II

DEFINITION OF THE MAPPING

§ 8. THE POSTULATES

Our object is to begin with a mapping I_0 defined on some "elementary" subset E of a partially ordered set F, and to extend the domain of definition to larger subsets with useful closure properties. This mapping I_0 is assumed to have a property related to, but weaker than, the property of being smoothly rising. The question is, should we assume "smoothly rising" or "σ-smoothly rising?" That is, should postulate (8.1e) below be a property of all directed sets, or only of countable directed sets? In any application, if the postulate can be shown to hold without the restriction to countable subsets, the resulting extension cannot lose, and may gain, in generality. On the other hand, there are cases in which it is difficult or impossible to establish the postulate for all directed sets, and not at all difficult when the restriction to countable sets is made. In such cases it is obviously desirable to develop the theory in the form in which the restriction is made, since the otherwise more desirable form is out of reach. Accordingly, we state both forms, a set (8.1) of postulates without restriction to countable sets, and a set (8.1σ) with the restriction. As before, we save space by using the convention of brackets; either all bracketed expressions are to be included, or else all are to be omitted.

Besides the partial ordering $\geqq$, we shall use a strengthening of it, named $\gg$. But whenever we use the symbol $\vee$ or $\wedge$, or the words bound or infimum or supremum, it shall be with reference to $\geqq$, never with reference to $\gg$.

(8.1 [σ]) POSTULATES.

(a) F is a [σ]-complete and infinitely distributive lattice under the partial ordering $\geqq$.

(b) $\gg$ is a strengthening of $\geqq$ (cf. (1.3)).

(c) G is a Dedekind complete partially ordered set, such that for each two elements g_1, g_2

of G, if g_1 and g_2 have an upper bound in G they also have a lower bound in G, and vice versa.

(d) I_o is an isotone function whose domain is a subset E of F and whose range is contained in G.

(e) For each pair S_1, S_2 of [countable] subsets of E directed by $\gg$, $\ll$ respectively and having $\bigvee S_1 \geqq \bigwedge S_2$, and such also that $\bigvee I_o(S_1)$ and $\bigwedge I_o(S_2)$ exist in G, the inequality $\bigvee I_o(S_1) \geqq \bigwedge I_o(S_2)$ holds.

(f) If e_1 and e_2 are in E, and $I_o(e_1)$ and $I_o(e_2)$ have a common upper or lower bound in G, there exist elements e' and e'' of E such that $e' \ll e_i \ll e''$, $i = 1, 2$.

(g) If e_1, e_2 and e_3 are in E, and $I_o(e_1)$ and $I_o(e_2)$ have a common upper or lower bound in G, then for every f in F such that $f \gg \text{mid}(e_1, e_2, e_3)$ there is an e in E such that $f \gg e \gg \text{mid}(e_1, e_2, e_3)$; and dually.

REMARK. (8.1c) holds when I is a partially ordered group, by (6.7). It clearly holds if G has a greatest and a least element, in particular when G is a complete lattice.

We now list some easy but useful consequences of the postulates.

(8.2) THEOREM. If (8.1 [σ]) holds:

(a) for each e in E, there exist e' and e'' in E such that $e' \ll e \ll e''$;

(b) if $e_1, \ldots, e_n$ are elements of E such that each pair $\{I_o(e_{i-1}), I_o(e_i)\}$ has an upper or a lower bound, there exist e' and e'' in E such that $e' \ll e_i \ll e''$ $(i = 1, \ldots, n)$;

(c) if $e_1, e_1', \ldots, e_n, e'_n$ are elements of E such that the $I_o(e_i)$ have an upper or lower bound, and $e_i \ll e'_i$ for $i = 1, \ldots, n$, there exists an e in E such that $\bigvee\{e_1, \ldots, e_n\} \ll e \ll \bigvee\{e'_1, \ldots, e'_n\}$.

(d) If e_1 and e_2 are in E and $I_o(e_1)$ and $I_o(e_2)$ have a common upper or lower bound, and f is an element of F such that $f \gg e_1$ and $f \gg e_2$, there exists an e_3 in F such that $f \gg e_3 \gg e_i$, $i = 1, 2$; and the set of three elements $I_0(e_1)$, $I_0(e_2)$, $I_0(e_3)$ is bounded. The dual is also true.

For (a) we take $e_1 = e_2 = e$ in (8.1f). By (8.1f) statement (b) holds for $n = 2$. Assume it true for $n = k - 1$, and let $e_1, \ldots, e_k$ satisfy the requirements of (b). By hypothesis there are elements e'_1, e''_1 such that $e'_1 \ll e_i \ll e''_1$ $(i = 1, \ldots, k-1)$, and there are elements e_2', e_2'' such that $e_2' \ll e_j \ll e_2''$ $(j = k-1, k)$. Now $I_0(e_1'')$ and $I_0(e_2'')$ have the common lower bound $I_0(e_{k-1})$, so by (8.1f) there exists an e'' in E such that $e'' \gg e_i''$, $I = 1, 2$. Hence $e'' \gg e_i$, $i = 1, \ldots, k$. The existence of a suitable e' is similarly proved, and by induction (b) is established.

To establish (d), by (8.1f) there exists an e_3 in E such that $e_3 \gg e_1$ and $e_3 \gg e_2$. Then $\text{mid}\,(e_1, e_2, e_3) = e_1 \vee e_2$, so (d) follows from (8.1g). From this we obtain (c) with $n = 2$; we need only choose $f = e_1' \vee e_2'$. Suppose (c) holds for $n = k - 1$, and let $e_1, \ldots, e_k'$ be as described in (c). Then there exists an e_0 in E such that $\bigvee\{e_1, \ldots, e_{k-1}\} \ll e_0 \ll \bigvee\{e_1', \ldots, e_{k-1}'\}$. By (d) there exists an e_0' in E such that $e_0 \ll e_0' \ll \bigvee\{e_1', \ldots, e_{k-1}'\}$. Since (c) holds for $n = 2$, there exists an e in E such that $e_0 \vee e_k \ll e \ll e_0' \vee e_k'$. Then $\bigvee\{e_1, \ldots, e_k\} \ll e \ll \bigvee\{e'_1, \ldots, e'k\}$, and by induction (c) is established.

If (8.2d) holds without the requirement of a common bound for $I_0(e_1)$ and $I_0(e_2)$, (8.1g) holds:

(8.3) **LEMMA**. If (8.1 [σ]a,b,c,d) hold; and for each pair of elements e_1, e_2 of E and each pair of elements f, f' of F such that $f \ll e_1 \vee e_2 \ll f'$ there exist elements e, e' of E such that $f \ll e \ll e_1 \vee e_2 \ll e' \ll f'$, and dually; then (8.1 [σ]g) holds in a strengthened form, the requirement of a common bound for $I_0(e_1)$ and $I_0(e_2)$ being superfluous.

For if $f \gg \text{mid}\,(e_1, e_2, e_3)$, then $f \gg e_1 \wedge e_2$, and likewise for the two other pairs. By hypothesis there exist e_1', e_2', e_3' in E such that

$$f \gg e_1' \gg e_2 \wedge e_3,\ f \gg e_2' \gg e_1 \wedge e_3,\ f \gg e_3' \gg e_1 \wedge e_2 .$$

Again by the hypothesis, there exists e_4 in E such that $f \gg e_4 \gg e_1' \vee e_2'$, and similarly there exists e_5 in E such that $f \gg e_5 \gg e_2' \vee e_3'$. By a third application of the hypothesis, there exists e in E such that $f \gg e \gg e_4 \vee e_5$. Then $e \gg e_1', e_2', e_3'$, so $e \gg \operatorname{mid}(e_1, e_2, e_3)$. This establishes half of the strengthened (8.1g); the other half is similarly proved.

§ 9. U-ELEMENTS AND L-ELEMENTS

The first step in the extension of domain of I_0, in the manner of Daniell, is to extend it to include two new subsets of F, which we shall call the sets of U-elements and of L-elements.

(9.1[σ]) DEFINITION. An element u of F is a U-element if there exists a [countable] subset S of E, directed by $\gg$, such that $\bigvee S = u$. Each such set S will be said to be associated with u. If such a set exists for which it is also true that $I_0(S)$ has an upper bound in G, u is a summable u-element.

An element l of F is an L-element if there exists a [countable] subset S of E, directed by $\ll$, such that $\bigwedge S = l$. Each such set will be said to be associated with l. If such a set exists for which it is also true that $I_0(S)$ has a lower bound in G, l is a summable L-element.

To be consistent, we should use some affix, preferably a σ, on the letters U and L to indicate whether we mean to use definitions (9.1) or (9.1σ). But to avoid complicating the notation we omit this affix. Consequently we must keep in mind that in all theorems in which (8.1) is a hypothesis, any U- or L-elements mentioned are assumed to satisfy (9.1), while if (8.1σ) is a hypothesis any U- or L-elements mentioned are assumed to satisfy (9.1σ).

(9.2) THEOREM. Let (8.1[σ]) hold. If S is a [countable] subset of E directed by $\gg$ and such that the set $I_0(S)$ has an upper bound, and e is in E and $e \ll \bigvee S$, then $I_0(e) \leqq \bigvee I_0(S)$.

By repeated application of (8.2d) we obtain a sequence $e_1, e_2, \ldots$ of elements of E such that $\bigvee S \gg e_1 \gg e_2 \gg \ldots \gg e$. Then $\bigvee I_0(S)$ and $\bigwedge\{I_0(e_j) : j = 1, 2, \ldots\}$ exist, and by (8.1[σ]d,e),

$\bigvee I_0(S) \geqq \bigwedge\{I_0(e_j) : j = 1, 2, \ldots\} \geqq I_0(e)$.

(9.3) COROLLARY. Let (8.1[σ]) hold. Let S_1 and S_2 be [countable] subsets of E directed by $\gg$, and let S_3 and S_4 be [countable] subsets of E directed by $\ll$. Then

(i) if $\bigvee S_1 \geqq \bigvee S_2$ and $\bigvee I_0(S_1)$ exists, $\bigvee I_0(S_2)$ also exists, and $\bigvee I_0(S_1) \geqq \bigvee I_0(S_2)$;

(ii) if $\bigvee S_1 \geqq \bigwedge S_3$, and $\bigvee I_0(S_1)$ and $\bigwedge I_0(S_3)$ both exist, then $\bigvee I_0(S_1) \geqq \bigwedge I_0(S_3)$;

(iii) if $\bigwedge S_3 \geqq \bigvee S_1$, then $\bigvee I_0(S_1)$ and $\bigwedge I_0(S_3)$ both exist, and $\bigwedge I_0(S_3) \geqq \bigvee I_0(S_1)$;

(iv) if $\bigwedge S_3 \geqq \bigwedge S_4$, and $\bigwedge I_0(S_4)$ exists, $\bigwedge I_0(S_3)$ also exists, and $\bigwedge I_0(S_3) \geqq \bigwedge I_0(S_4)$.

Under the assumption in (i), if e_2 is in S_2, by hypothesis there is an e_2' in S_2 such that $e_2 \ll e_2'$, so $e_2 \ll \bigvee S_2 \leqq \bigvee S_1$. By (9.2), $I_0(e_2) \leqq \bigvee I_0(S_1)$. This holds for all e_2 in S_2, so by (8.1[σ]c) the conclusion is established. Dually, (iv) holds; and (ii) is merely a re-statement of (8.1e). To establish (iii), let e_1 belong to S_1 and e_3 to S_3. Then $e_3 \geqq \bigwedge S_3 \geqq \bigvee S_1 \geqq e_1$, so $I_0(e_3) \leqq I_0(e_1)$, whence (iii) follows.

(9.4) COROLLARY AND DEFINITION. Let (8.1[σ]) hold. If u is a summable U-element, for all sets S associated with u the supremum $\bigvee I_0(S)$ has the same value. This common value is denoted by the symbol $I_1(u)$. Dually, if l is a summable L-element, for all sets S associated with l the infimum $\bigwedge I_0(S)$ has the same value. This common value is denoted by the symbol $I_1(l)$.

The assertions concerning u follow from (9.3i) and those concerning l from (9.3iv).

It would seem that the symbol $I_1(f)$ could be ambiguous, since f might be both a summable U-element and a summable L-element. (In fact, any f which is both a U-element and an L-element is easily seen to be a summable U-element and a summable L-element.) However, by (9.3ii,iii), when this happens the value found for $I_1(f)$ when f is regarded as a summable U-element is the same as the value found when f is regarded as a summable L-element, so the ambiguity does not cause trouble.

(9.5) COROLLARY. If f_1 is a summable U- or L-element and f_2 is a summable U- or L-element, and $f_1 \leqq f_2$, then $I_1(f_1) \leqq I_1(f_2)$.

This follows at once from (9.3) and (9.4).

(9.6) COROLLARY. If u is a summable U-element, and e is in E, and $e \ll u$, then $I_0(e) \leqq I_1(u)$.

This follows from (9.2) and (9.4).

(9.7) LEMMA. Let (8.1[σ]) hold. If e is in E, there exist summable U-elements u_1, u_2 and summable L-elements l_1, l_2 such that $u_1 \ll e \ll u_2$, $l_1 \ll e \ll l_2$. If e_1 and e_2 are in E and $e_1 \ll e_2$, there exist a summable U-element u and a summable L-element l such that $e_1 \ll l \ll e_2$, $e_1 \ll u \ll e_2$.

We prove the second statement first. By repeated use of (8.2d), we obtain elements $e_3, e_4, e_5, \ldots$ of E such that $e_1 \ll e_4 \ll e_5 \ll e_6 \ll \ldots \ll e_3 \ll e_2$. Then $\bigvee\{e_n : n = 4, 5, \ldots\}$ is a U-element $\leqq e_3 \ll e_2$. It is clearly summable and $\gg e_1$. The L-element l is similarly defined. The first conclusion follows from the second, with (8.2a).

§ 10. LATTICE PROPERTIES OF THE CLASSES OF U- AND L-ELEMENTS

In several theorems we shall make use of the same construction, which we therefore isolate and state.

(10.1) CONSTRUCTION. Let (8.1[σ]) be satisfied. If $S_1, \ldots, S_n$ are [countable] subsets of E directed by $\gg$, and for each set $e_1, \ldots, e_n$ with e_i in S_i $(i = 1, \ldots, n)$ the elements $I_0(e_1), \ldots, I_0(e_n)$ have a common upper or lower bound, sets S_{sup} and S_{inf} can be constructed as follows. Form all 2n-uples $e_1, e_1', \ldots, e_n, e_n'$ with e_i and e_i' in S_i and $e_i \ll e_i'$, $i = 1, \ldots, n$. To each such

2n-uple corresponds by (8.2c) at least one element e' in E such that $\bigvee(e_1, \ldots, e_n) \ll e' \ll \bigvee(e_1', \ldots, e_n')$ and at least one element e'' in E such that $\bigwedge(e_1, \ldots, e_n) \ll e'' \ll \bigwedge(e_1', \ldots, e_n')$. For each such 2n-uple we choose exactly one e' and exactly one e'' with the property stated. The aggregate of chosen e' constitutes the set S_{sup}; the aggregate of chosen e'' constitutes the set S_{inf}.

Likewise, if S_1, S_2, S_3 are subsets of E directed by $\gg$, and for each e_1 in S_1 there exists an e_2 in S_2 such that $e_2 \gg e_1$, a set S_{mid} can be constructed as follows. Form all sextuples $e_1, e_1', e_2, e_2', e_3, e_3'$ with e_i and e_i' in S_i, $e_i \ll e_i$ $(i = 1, 2, 3)$ and $e_1 \ll e_2$ and $e_1' \ll e_2'$. To each such sextuple corresponds by (8.1[σ],g,b) an element e of E such that $\text{mid}\,(e_1, e_2, e_3) \ll e \ll \text{mid}\,(e_1', e_2', e_3')$. For each such sextuple we choose exactly one e as described; the aggregate of chosen elements e is the set S_{mid}.

The following properties are easily established.

(10.2) COROLLARY. With the notation of (10.1),

(a) if $S_1, \ldots, S_n$ are countable so are S_{sup} and S_{inf}, and if S_1, S_2 and S_3 are countable, so is S_{mid};

(b) S_{sup}, S_{inf} and S_{mid} are directed by $\gg$;

(c) $\bigvee S_{sup} = \bigvee(\bigvee S_1, \ldots, \bigvee S_n)$, $\bigvee S_{inf} = \bigwedge(\bigvee S_1, \ldots, \bigvee S_n)$, $\bigvee S_{mid} = \text{mid}\,(\bigvee S_1, \bigvee S_2, \bigvee S_3)$.

Define S' to be the set of all elements $\{\bigvee(e_1, \ldots, e_n)$: e_1 in $S_1, \ldots, e_n$ in $S_n\}$, and S'' to be the set $\{\bigwedge(e_1, \ldots, e_n)$: e_1 in $S_1, \ldots, e_n$ in S_n.$\}$ If e is in S' it has the form $\bigvee(e_1, \ldots, e_n)$ for some e_1 in $S_1, \ldots, e_n$ in S_n, so by (10.1) there is an e' in S_{sup} such that $e' \geqq e$. Conversely, if e' is in S_{sup}, there are elements $e_1', \ldots, e_n'$ in $S_1, \ldots, S_n$ respectively such that $\bigvee(e_1', \ldots, e_n') \gg e'$; and the left member of this inequality is in S'. Hence $\bigvee S' = \bigvee S_{sup}$. But by (2.8) this implies the first equation in (c). The proofs of the other two are similar, the reference to (2.8) being replaced by references to (2.9) and (2.14) respectively.

To establish (b), we note that since each S_i is directed by $\gg$, so are S' and S''. If e_1 and e_2 are in S_{sup}, there are elements e_1', e_2' in S' such that $e_1' \gg e_1$ and $e_2' \gg e_2$; S' being directed, there is an element e' in S' such that $e' \gg e_1'$ and $e' \gg e_2'$; and finally there is an e in S_{sup} such that $e \gg e'$. The proofs of the other statements in (b) are similar.

(10.3) COROLLARY. Let (8.1[σ]) hold. If $u_1, \ldots, u_n$ are summable U-elements, and $u_1 \geqq u_i$ $(i = 2, \ldots, n)$, and $S_2, \ldots, S_n$ are associated with $u_2, \ldots, u_n$ respectively, there exists a set S associated with u_1 such that if $e_2, \ldots, e_n$ are in $S_2, \ldots, S_n$ respectively, there is an e in S such that $e \gg e_i$, $i = 2, \ldots, n$.

Let S_1 be associated with u_1. Then $I_1(u_1)$ is an upper bound for $I_0(e_i)$ for all e_i in S_i, $i = 1, \ldots, n$, and (10.1) applies. The set $S = S_{sup}$ has the desired properties.

(10.4) COROLLARY. Let (8.1[σ]) hold. If u is a summable U-element and e is in E and $e \ll u$, there exists a set S associated with u and such that for every e' in S it is true that $e' \gg e$.

Let S_1 be any set associated with u. By repeated use of (8.2d) we select a set S_2 of elements $e_1, e_2, \ldots$ of E such that $e \ll e_1 \ll e_2 \ll \ldots \ll u$. Then $\bigvee S_2 \leqq u$. Let S be the set S_{sup} formed by (10.1) from S_1 and S_2; then $e' \gg e$ for all e' in S, and $\bigvee S = \bigvee\{\bigvee S_1, \bigvee S_2\} = u$.

(10.5) THEOREM. Let (8.1[σ]) hold. If $u_1, \ldots, u_n$ are summable U-elements and $I_1(u_1), \ldots, I_1(u_n)$ have a common upper or lower bound, $\bigwedge\{u_1, \ldots, u_n\}$ is a summable U-element.

For each u_j, let $S(u_j)$ be associated with u_j, $j = 1, \ldots, n$. By (8.1c) and (6.7), the elements $I_1(u_1), \ldots, I_1(u_n)$ have an upper bound, which is also an upper bound for $I_0(e_j)$, e_j in S_j, $j = 1, \ldots, n$. So construction (10.1) can be applied to define S_{inf}. By (10.2), $\bigvee S_{inf} = \bigwedge\{u_1, \ldots, u_n\}$, so this is a U-element. Being $\leqq u_1$, it is summable.

(10.6) THEOREM. Let K be a [countable] collection of

summable U-elements such that for each pair u_1, u_2 of elements of K, $I_1(u_1)$ and $I_1(u_2)$ have a common upper or lower bound. Then $\bigvee K$ is a U-element.

For each u in K, let S(u) be associated with u. For each finite subset $u_1, \ldots, u_n$ of elements of K, by (8.1c) and (6.7) we see that $I_1(u_1), \ldots, I_1(u_n)$ have an upper bound. So we can apply (10.1) to the sets $S(u_1), \ldots, S(u_n)$ to construct S_{sup}. Let S be the union of all the sets S_{sup} thus constructed. [Then S is countable]. If e_1 and e_1' are in S, we can find two finite sets $u_1, \ldots, u_n$ and $u_1', \ldots, u_m'$ of elements of K and elements $e_1, \ldots, e_m'$ of $S(u_1), \ldots, S(u_m')$ respectively such that $e \ll \bigvee\{e_1, \ldots, e_n\}$ and $e' \ll \bigvee\{e_1', \ldots, e_m'\}$. If any u_i is also an u_j', we can choose the same element as e_i and e_j', since the S[u] are directed by $\gg$. But now S contains an element $e'' \gg e_1, \ldots, e_n, e_1', \ldots, e_m'$, so S is directed by $\gg$. Therefore $\bigvee S$ is a U-element. If e is in S(u), there is an e' in S such that $e' \gg e$, so $\bigvee S \geqq e$, whence $\bigvee S \geqq \bigvee S(u) = u$. This holds for all u in K, so $\bigvee S \geqq \bigvee K$. On the other hand, for each e in S there are sets $S(u_1), \ldots, S(u_n)$ and members $e_1, \ldots, e_n$ of those sets such that $e \ll \bigvee\{e_1, \ldots, e_n\} \leqq \bigvee\{u_1, \ldots, u_n\} \leqq \bigvee K$. Hence $\bigvee S \leqq \bigvee K$. This proves equality, so $\bigvee K$ is equal to the U-element $\bigvee S$.

(10.7) **THEOREM.** Let K be a [countable] collection of summable U-elements directed by $\geqq$. Then $\bigvee K$ is a summable U-element if and only if the set $I_1(K)$ has an upper bound in G; and in that case $I_1(\bigvee K) = \bigvee I_1(K)$.

If $\bigvee K$ is summable, $I_1(\bigvee K)$ is obviously an upper bound for $I_1(u)$, u in K. So

$$I_1(\bigvee K) \geqq \bigvee I_1(K) .$$

Conversely, suppose there is an upper bound for $I_1(K)$. Using the notation of (10.6), we construct the set S. For each e in S there are elements $u_1, \ldots, u_n$ of K such that $e \ll \bigvee\{u_1, \ldots, u_n\}$; and since K is directed, there is an element u of K such that $u \geqq \bigvee\{u_1, \ldots, u_n\}$. Hence $e \ll u$, and $I_0(e) \leqq I_1(u) \leqq \bigvee I_1(K)$. Then by definition $\bigvee K = \bigvee S$ is a summable U-element, and also

$$I_1(\bigvee K) = I_1(\bigvee S) = \bigvee\{I_0(e) : e \text{ in } S\}$$
$$\leqq \bigvee I_1(K) .$$

The reverse inequality was established above, so equality holds.

(10.8) THEOREM. Let (8.[σ]) hold. If u_1, u_2 and u_3 are U-elements and $u_1 \leqq u_2$, and u_2 is summable, mid (u_1,u_2,u_3) is a summable U-element.

Let S_1 and S_3 be sets associated with u_1 and u_3 respectively. By (10.3), there is a [countable] subset S_2 of E directed by $\gg$ having $\bigvee S_2 = u_2$, and such that for each e_1 in S_1 there is an e_2 in S_2 for which $e_2 \gg e_1$. Then by (10.1) we can construct S_{mid}. This is [countable and] directed by $\gg$, and $\bigvee S_{mid} = \text{mid}\,(u_1,u_2,u_3)$ by (10.2). So mid (u_1,u_2,u_3) is a U-element. Since it is $\leqq u_2$, it is a summable U-element.

§ 11. DEFINITION OF THE MAPPING OR INTEGRAL

Again we shall assume (8.1) or (8.1σ) to be satisfied. We shall frequently need to refer to the set of all summable U-elements above a given f of F, so we shall provide a symbol for this set, and also for its dual.

(11.1) DEFINITION. If f is in F, U $[\geqq f]$ is the class of all summable U-elements u satisfying $u \geqq f$, and $L[\leqq f]$ is the class of all summable L-elements l satisfying $l \leqq f$.

(11.2) LEMMA. Let (8.1[σ]) hold. If f is in F and $L[\leqq f]$ is not empty, the set of elements of G of the form $I_1(u)$(u in $U[\geqq f]$) is directed by $\leqq$; and dually.

If g_1 and g_2 are in the class, there are elements u_1,u_2 of $U[\geqq f]$ such that $g_1 = I_1(u_1)$ and $g_2 = I_1(u_2)$. For any l in $L[\leqq f]$, $I_1(l) \leqq g_1,g_2$. So by (10.5), $u_1 \wedge u_2$ is a summable U-element, and it is clearly $\geqq f$, so it is in $U[\geqq f]$. Also, by (9.5)

$$I_1(u_1 \wedge u_2) \leqq I_1(u_i) = g_i \qquad (i = 1, 2) .$$

(11.3) DEFINITION. If f is an element of F such that both $U[\geqq f]$ and $L[\leqq f]$ are non-empty, we define the upper and lower integrals [or images] of f by the respective formulas

$$\overline{I}(f) = \bigwedge I_1(U[\geqq f]) ,$$

$$\underline{I}(f) = \bigvee I_1(L[\leqq f]) .$$

These surely exist; for if u' is in $U[\geqq f]$ and l' in $L[\leqq f]$, then $I_1(u')$ is an upper bound for $I_1(L[\leqq f])$ and $I_1(l')$ is a lower bound for $I_1(U[\geqq f])$, and by (11.2) and (8.1c) the right members of the above equations exist.

If $U[\geqq f]$ and $L[\leqq f]$ are non-empty, each element u of the former is $\geqq$ each element l of the latter, so by (9.5) $I_1(u) \geqq I_1(l)$. Thus $I_1(u)$ is an upper bound for $I_1(L[\leqq f])$, and must be $\geqq$ its supremum $\underline{I}(f)$. Then $I_1(u) \geqq \underline{I}(f)$, and $\underline{I}(f)$ is a lower bound for $I_1(U[\geqq f])$, so must be $\leqq$ its infimum, $\overline{I}(f)$. So:

(11.4) THEOREM. Let (8.1[σ]) hold. If f is in F, and $\overline{I}(f)$ and $\underline{I}(f)$ exist, they satisfy the relation $\underline{I}(f) \leqq \overline{I}(f)$.

Another corollary of the definition is the following:

(11.5) THEOREM. Let (8.1[σ]) hold. If f_1 and f_2 are in F and both have upper and lower integrals, and $f_1 \leqq f_2$, then $\overline{I}(f_1) \leqq \overline{I}(f_2)$ and $\underline{I}(f_1) \leqq \underline{I}(f_1)$.

For then $U[\geqq f_1] \supset U[\geqq f_2]$, so $I_1(U[\geqq f_1]) \supset I_1(U[\geqq f_2])$, and the former set has the smaller infimum, that is $\overline{I}(f_1) \leqq \overline{I}(f_2)$. The other conclusion is established analogously.

(11.6) DEFINITION. Let (8.1[σ]) hold. If f is an element of F for which $\overline{I}(f)$ and $\underline{I}(f)$ are defined and equal, we define the integral, or image, of f to be $I(f) = \underline{I}(f) = \overline{I}(f)$. In this case f is called a summable element.

We have already used the adjective "summable" in (9.1); so we shall show that there is no contradiction between the two uses.

(11.7) THEOREM. Let (8.1[σ]) hold. If f is in F and is a U- or L-element, it is a summable U- or L-element as defined in (9.1) if and only if it is summable as defined in (11.6). In this case $I(f) = I_1(f)$.

We consider U-elements only; the case of L-elements can be

discussed dually. Suppose first that f is a U-element summable in the sense of (11.6). Then there is a set S associated with f, and there exists a U-element u summable in the sense of (9.1), such that $f \leqq u$. For all e in S we have $e \ll \bigvee S = f \leqq u$, so by (9.6) $I_0(e) \leqq I_1(u)$. Thus $I_0(S)$ has an upper bound in G, and f is a summable U-element in the sense of (9.1).

Conversely, suppose that f is a summable U-element, as defined in (9.1). Let S be associated with f. If e_0 is in S then $e_0 \ll f$, so by (9.7) there is a summable L-element $\ll e_0$, and $L[\leqq f]$ is not empty; and $U[\geqq f]$ contains f itself, so is not empty. Also, if e_0 is in S, by (9.7) there is a summable L-element l such that $e_0 \ll l \ll f$, so $I_0(e_0) \leqq \bigvee I_1(L[\leqq f]) = \underline{I}(f)$. This holds for all e_0 in S, so

$$I_1(f) = \bigvee I_0(S) \leqq \underline{I}(f) \ .$$

On the other hand, since f is in $U[\geqq f]$,

$$\overline{I}(f) = \bigwedge I_1(U[\geqq f]) \leqq I_1(f) \ .$$

Combining these inequalities with (11.4), we obtain $\overline{I}(f) = \underline{I}(f) = I_1(f)$, completing the proof.

The following is a trivial corollary of (11.5).

(11.8) COROLLARY. Let (8.1[σ]) hold. If f_1 and f_2 are summable, and $f_1 \leqq f_2$, then $I(f_1) \leqq I(f_2)$.

Another easy corollary is

(11.9) COROLLARY. Let (8.1[σ]) hold. If f is summable, so are $f_1 = \bigwedge U[\geqq f]$ and $f_2 = \bigvee L[\leqq f]$, and $I(f) = I(f_1) = I(f_2)$.

CHAPTER III

LATTICE PROPERTIES, CONVERGENCE AND MEASURABILITY

§ 12. FURTHER LATTICE PROPERTIES OF THE CLASSES OF U-ELEMENTS AND L-ELEMENTS

In order to establish lattice properties of summable elements it is apparently necessary to strengthen the postulates (8.1). We do this in two ways. First, we assume that G is an additive group; this is not a great restriction, as the applications show. Second, we assume, roughly stated, that in $I_0(\text{mid}\ (e_1,e_2,e_3))$ if we increase e_1,e_2 and e_3 so that each $I_0(e_j)$ has a small increase, then $I_0(\text{mid}\ (e_1,e_2,e_3))$ also has a small increase, not more than the sum of the increases of the $I_0(e_j)$. But the statement of this property is a trifle involved, because we have not assumed that $\text{mid}\ (e_1,e_2,e_3)$ is in E.

(12.1 [σ]) POSTULATE.

(a) Same as (8.1[σ]a).

(b) Same as (8.1[σ]b).

(c) G is a Dedekind complete partially ordered group under a commutative operation +.

(d), (e), (f), (g) Same as (8.1[σ]d,e,f,g) respectively.

(h) If $e_1,e_2,e_3,e_1',e_2',e_3',e_4$ and e_5 are members of E such that $e_i \leqq e_i'$ (i = 1, 2, 3), $e_1 \leqq e_2$, $e_1' \leqq e_2'$, and

$$\text{mid}\ (e_1,e_2,e_3) \leqq e_k \leqq \text{mid}\ (e_1',e_2',e_3') \qquad (k = 4, 5)\ ,$$

then

$$I_0(e_5) - I_0(e_4) \leqq \sum_{j=1}^{3} [I_0(e_j') - I_0(e_j)]\ .$$

As remarked after (8.1), (12.[σ]c) implies (8.1[σ]c), so (12.1[σ]) implies (8.1[σ]). The new postulate is itself a consequence of more familiar assumptions, but these we postpone until a more appropriate time.

(12.2) **LEMMA**. Let (12.1[σ]) hold. Let e_1, e_2, e_1', e_2' be elements of E such that $e_1 \ll e_1'$, $e_2 \ll e_2'$, and $I_0(e_1)$ and $I_0(e_2)$ have a common upper or lower bound. Let e_4, e_5 be elements of E which satisfy $e_1 \vee e_2 \ll e_k \ll e_1' \vee e_2'$ $(k = 4, 5)$, or else which satisfy $e_1 \wedge e_2 \ll e_k \ll e_1' \wedge e_2'$ $(k = 4, 5)$. Then

$$I_0(e_5) - I_0(e_4) \leqq \sum_{j=1}^{2} [I_0(e_j') - I_0(e_j)] .$$

By (8.2b), there are elements e', e'' in E such that $e' \ll e_j \ll e''$ and $e' \ll e_j' \ll e''$, $(j = 1, 2)$. Choose $e_3 = e_3' = e''$; then mid $(e_1, e_2, e_3) = e_1 \vee e_2$ and mid $(e_1', e_2', e_3') = e_1' \vee e_2'$. Hence under the first assumption on e_4 and e_5 we have mid $(e_1, e_2, e_3) \leqq e_k \leqq$ mid (e_1', e_2', e_3'). Also, by interchanging the subscripts 2 and 3 we cause the inequalities $e_1 \leqq e_2$ and $e_1' \leqq e_2'$ to be satisfied. By (12.1[σ]h), the conclusion of (12.2) holds. A similar proof, using e' instead of e'', establishes the conclusion under the alternative hypothesis.

We here digress slightly to show that under certain circumstances the converse of (12.2) holds.

(12.3) **LEMMA**. If (12.1[σ]a,b,c,d,e,f,g) hold, and E is a lattice under $\geqq$, and the conclusion of (12.2) holds for all sets $e_1, e_1', e_2, e_2', e_4, e_5$ as described, then (12.1[σ]h) holds.

Let $e_1, \ldots, e_3', e_4$ and e_5 be as in (12.1[σ]h). Then

$$\begin{aligned} I_0(e_5) - I_0(e_4) &\leqq I_0(\text{mid } (e_1', e_2', e_3')) - I_0(\text{mid } (e_1, e_2, e_3)) \\ &= I_0(e_1' \vee [e_2' \wedge e_3']) - I_0(e_1 \vee [e_2 \wedge e_3]) \\ &\leqq [I_0(e_1') - I_0(e_1)] + [I_0(e_2' \wedge e_3') - I_0(e_2 \wedge e_3)] \\ &\leqq \sum_{j=1}^{3} [I_0(e_j') - I_0(e_j)] , \end{aligned}$$

and (12.1[σ]h) is satisfied.

Under (12.1[σ]) we can sharpen the results obtained in (10.6) as follows.

(12.4) **THEOREM**. Let (12.1[σ]) hold, and let u_1 and u_2 be summable U-elements such that $I_1(u_1)$ and $I_1(u_2)$

have a common upper or lower bound. Then $u_1 \vee u_2$ is a summable U-element.

Let S_1 be associated with u_1 and S_2 with u_2. We form S_{sup} as in (10.1). Let e^* be a fixed element of S_{sup}, and e_1^* and e_2^* elements of S_1, S_2 respectively, such that $e^* \gg e_1^* \vee e_2^*$. Let e be an arbitrary member of S_{sup}. There exists an $\bar{e}$ in S_{sup} such that $\bar{e} \gg e$ and $\bar{e} \gg e^*$. There exist an e_1' in S_1 and an e_2' in S_2 such that $\bar{e} \ll e_1' \vee e_2'$, and without loss of generality we may assume that $e_i' \gg e_i^*$, $i = 1, 2$. Then by (12.1)

$$\begin{aligned} I_0(e) &\leqq I_0(\bar{e}) \\ &\leqq I_0(e^*) + \sum_{j=1}^{2} [I_0(e_j') - I_0(e_j^*)] \\ &\leqq I_0(e^*) + \sum_{j=1}^{2} [I_1(u_j) - I_0(e_j^*)] , \end{aligned}$$

so $I_0(S_{sup})$ is bounded above, and $u_1 \vee u_2 = \vee S_{sup}$ is summable.

The next theorem is in effect an extension of (12.1[σ]h).

(12.5) **THEOREM**. Let (12.1[σ]) hold. Let f_1, f_2, f_3 be three summable U-elements or three summable L-elements, and let f_1', f_2', f_3' also be three summable U-elements or three summable L-elements. Assume that $f_1 \leqq f_2$, $f_1' \leqq f_2'$, and $f_i \leqq f_i'$, $i = 1, 2, 3$. Then

$$I_1(\text{mid}\,(f_1', f_2', f_3')) - I_1(\text{mid}\,(f_1, f_2, f_3)) \leqq \sum_{i=1}^{3} [I_1(f_i') - I_1(f_i)] .$$

The proof of the case in which the f_i are U-elements and the f_i' are L-elements will be omitted, since this situation will not arise in later proofs. Suppose first that the f_i and f_i' are all U-elements. Let S_1 and S_3 be [countable] subsets of E directed by $\gg$ and having $\vee S_1 = f_1$, $\vee S_3 = f_3$. By (10.3) there are [countable] subsets S_1', S_2 and S_3' of E directed by $\gg$, having $\vee S_1' = f_1'$, $\vee S_2 = f_2$ and $\vee S_3' = f_3'$, and such also that if e_1 is in S_1 and e_3 in S_3, there are elements e_1', e_2, e_3' of S_1, S_2, S_3 respectively for which $e_1' \gg e_1, e_2 \gg e_1$ and $e_3' \gg e_3$. Also there is a [countable] subset S_2' of E directed by $\gg$, having $S_2' = f_2'$, and such that if e_2 is in S_2 and e_1' in S_1', there exists an e_2' in S_2' for which $e_2' \gg e_2$ and $e_2' \gg e_1'$. Now (10.1) can be applied to the sets S_1, S_2, S_3 to form S_{mid}, and to S_1', S_2', S_3' to form S_{mid}'.

Let e_1, e_2 and e_3 be any members of S_1, S_2 and S_3 respectively such that $e_2 \gg e_1$. For some e in S_{mid} we have $e \gg \text{mid}\,(e_1,e_2,e_3)$. Let e' be any member of S'_{mid}. There exist elements e_1',e_2',e_3' belonging respectively to S_1,S_2,S_3 for which $\text{mid}\,(e_1',e_2',e_3') \gg e'$. We may assume that $e_i' \gg e_i$, $i = 1, 2, 3$, and also that $e_2' \gg e_1'$. Then by (12.1h)

$$I_0(e') - I_0(e) \leqq \sum_{i=1}^{3} (I_0(e_i') - I_0(e_i)) ,$$

whence

$$I_0(e') - I_1(\text{mid}\,(u_1,u_2,u_3)) \leqq \sum_{i=1}^{3} (I_1(u_i') - I_0(e_i)) .$$

Because e',e_1,e_2,e_3 are arbitrary members of S'_{mid},S_1,S_2,S_3 respectively, by (6.8) and (9.4) this inequality implies the conclusion of the theorem.

The case in which all six elements are L-elements is discussed similarly.

Now let f_1',f_2' and f_3' be U-elements, while f_1,f_2,f_3 are L-elements. Let L_i be the class of L-elements λ_i satisfying the inequality $f_i \leqq \lambda_i \leqq f_i'$, $(i = 1, 2, 3)$. We first establish an auxiliary statement.

$$(*) \qquad I_1(\text{mid}\,(f_1',f_2',f_3')) = \bigvee\{I_1(\text{mid}\,(\lambda_1,\lambda_2,\lambda_3)):\lambda_i \text{ in } L_i ,\ i = 1, 2, 3\} .$$

For each such $\lambda_1,\lambda_2,\lambda_3$ we have $\text{mid}\,(\lambda_1,\lambda_2,\lambda_3) \leqq \text{mid}\,(f_1',f_2',f_3')$, so by (9.5) the left member of (*) is an upper bound for $I_1(\text{mid}\,(\lambda_1,\lambda_2,\lambda_3))$ with λ_i in L_i. Let S_1',S_2',S_3' be [countable] subsets of E directed by $\gg$ and having $\bigvee S_i' = f_i'$, $i = 1, 2, 3$. Form S'_{mid} as in (10.1). If e is in S'_{mid}, there exist elements e_1,e_2,e_3 in S_1',S_2',S_3' respectively such that $e \ll \text{mid}\,(e_1,e_2,e_3)$. Also there exist e_i' in S_i' satisfying $e_i' \gg e_i$, $i = 1, 2, 3$. By (9.7) there is an L-element l_i satisfying $e_i \ll l_i \ll e_i'$, $i = 1, 2, 3$. Let $\lambda_i = l_i \vee f_i$, $i = 1, 2, 3$. Then λ_i is in L_i. Also,

$$\begin{aligned} \text{mid}\,(\lambda_1,\lambda_2,\lambda_3) &\geqq \text{mid}\,(l_1,l_2,l_3) \\ &\gg \text{mid}\,(e_1,e_2,e_3) \\ &\gg e , \end{aligned}$$

so $I_1(\text{mid}\,(\lambda_1,\lambda_2,\lambda_3)) \geqq I_0(e)$. Thus every upper bound for the set $\{I_1(\text{mid}\,(\lambda_1,\lambda_2,\lambda_3)):\lambda_i \text{ in } L_i,\ i = 1, 2, 3)\}$ is also an upper bound for

$I_0(e)$, e being an arbitrary member of S'_{mid}. It is then also an upper bound for $\bigvee\{I_0(e)$: e in $S'_{mid}\}$, which is $I_1(\text{mid}\,(f_1', f_2', f_3'))$. This completes the proof of (*).

By the part of the proof already completed, for each $\lambda_1, \lambda_2, \lambda_3$ with λ_i in L_i we have

$$I_1(\text{mid}\,(\lambda_1, \lambda_2, \lambda_3)) - I_1(f_1, f_2, f_3)$$
$$\leqq \sum_{i=1}^{3} (I_1(\lambda_i) - I_1(f_i))$$
$$\leqq \sum_{i=1}^{3} (I_1(f_i') - I_1(f_i)) .$$

This, with (*), completes the proof.

(12.6) COROLLARY. Let $(12.1[\sigma])$ hold. Let f_1 and f_2 both be summable u-elements or both be summable L-elements, and let f_1' and f_2' also both be summable U-elements or summable L-elements. Assume that $f_1 \leqq f_1'$ and $f_2 \leqq f_2'$, and also that $I_1(f_1)$ and $I_1(f_2)$ have a common upper or lower bound. Then

$$I_1(f_1' \vee f_2') - I_1(f_1 \vee f_2) \leqq \sum_{i=1}^{2} [I_1(f_i') - I_1(f_i)]$$

and

$$I_1(f_1' \wedge f_2') - I_1(f_1 \wedge f_2) \leqq \sum_{i=1}^{2} [I_1(f_i') - I_1(f_i)] .$$

By (10.5) and (12.4), if f_1 and f_2 are summable U-elements so are $f_1 \vee f_2$ and $f_1 \wedge f_2$, and if they are summable L-elements so are $f_1 \vee f_2$ and $f_1 \wedge f_2$. If in (12.5) we choose $f_1, f_2, f_3, f_1', f_2', f_3'$ to be the present $f_1, f_1' \vee f_2', f_2, f_1', f_1' \vee f_2', f_2'$ respectively we obtain the first inequality in our conclusion. If we choose them to be the present $f_1 \wedge f_2, f_1, f_2, f_1 \wedge f_2, f_1', f_2'$ respectively we obtain the second inequality.

All the examples in our last chapter satisfy a condition stronger than (12.2), as follows.

(12.7) POSTULATE. Let e_1, e_2, e_3, e_4 be elements of E such that $I_0(e_1)$ and $I_0(e_2)$ have a common upper or lower bound. Then: if $e_3 \ll e_1 \vee e_2$ and $e_4 \ll e_1 \wedge e_2$, then

$$I_0(e_3) + I_0(e_4) \leqq I_0(e_1) + I_0(e_2) ,$$

and if $e_3 \gg e_1 \vee e_2$ and $e_4 \gg e_1 \wedge e_2$, then

$$I_0(e_3) + I_0(e_4) \geqq I_0(e_1) + I_0(e_2) .$$

(12.8) **LEMMA.** If (12.1[σ]) and (12.7) hold, and u_1 and u_2 are summable U-elements such that $I_1(u_1)$ and $I_1(u_2)$ have a common upper or lower bound, then $u_1 \vee u_2$ and $u_1 \wedge u_2$ are summable, and $I_1(u_1 \vee u_2) + I_1(u_1 \wedge u_2) = I_1(u_1) + I_1(u_2)$.

Let S_1 be associated with u_1 and S_2 with u_2. We form S_{sup} and S_{inf} as in (10.1). Given any e' in S_{sup} and e'' in S_{inf}, we can find e_1 in S_1 and e_2 in S_2 such that $e' \ll e_1 \vee e_2$ and $e'' \ll e_1 \wedge e_2$. Hence by (12.7)

$$\begin{aligned} I_0(e') + I_0(e'') &\leqq I_0(e_1) + I_0(e_2) \\ &\leqq I_1(u_1) + I_1(u_2) , \end{aligned}$$

and by this and (6.1), $I_1(u_1 \vee u_2) + I_1(u_1 \wedge u_2) = \vee I_0(S_{sup}) + \vee I_0(S_{inf}) \leqq I_1(u_1) + I_1(u_2)$. On the other hand, if e_1 is in S_1 and e_2 in S_2, there exist elements e' in S_{sup} and e'' in S_{inf} such that $e' \gg e_1 \vee e_2$ and $e'' \gg e_1 \wedge e_2$, whence

$$I_0(e_1) + I_0(e'') \geqq I_0(e_1) + I_0(e_2) .$$

Therefore $I_1(u_1 \vee u_2) + I_1(u_1 \wedge u_2) \geqq I_0(e_1) + I_0(e_2)$, whence by (6.1) $I_1(u_1 \vee u_2) + I_1(u_1 \wedge u_2) \geqq I_1(u_1) + I_1(u_2)$. This completes the proof.

§ 13. LATTICE PROPERTIES OF THE CLASS OF SUMMABLE ELEMENTS

Postulates (12.1[σ]) are not sufficient to cause the class of summable elements to be a sublattice of F. However, we can establish some closely related properties.

(13.1) **THEOREM.** Let (12.1[σ]) hold. If f_1 and f_2 are any two summable elements of F such that $I(f_1)$ and $I(f_2)$ have a common bound, $f_1 \vee f_2$ and $f_1 \wedge f_2$ are summable, and for every summable element f_3 of

F, mid (f_1,f_2,f_3) is also summable.

Let us first suppose that f_1,f_2 and f_3 are summable and that $f_1 \leqq f_2$. Let u_1,u_2,u_3 be summable U-elements such that $u_i \geqq f_i$, $i = 1, 2, 3$. By (10.5), $u_1 \wedge u_2$ is also a summable U-element, and $u_1 \wedge u_2 \geqq f_1$. By (10.8), mid $(u_1 \wedge u_2,u_2,u_3)$ is a summable U-element, and it is clearly $\geqq$ mid (f_1,f_2,f_3). Likewise, if l_i is in $L[\leqq f_i]$, $i = 1, 2, 3$, mid $(l_1,l_1 \vee l_2,l_3)$ is a summable L-element and is $\geqq$ mid (f_1,f_2,f_3). Thus the upper and lower integrals of mid (f_1,f_2,f_3) exist, and by (12.5)

$$\overline{I}(\text{mid }(f_1,f_2,f_3)) - \underline{I}(\text{mid }(f_1,f_2,f_3))$$

$$\leqq I_1(\text{mid }(u_1 \wedge u_2,u_2,u_3))$$

$$- I_1(\text{mid }(l_1,l_1 \vee l_2,l_3))$$

$$\leqq \sum_{i=1}^{3} [I_1(u_i) - I_1(l_i)] .$$

Each of the three terms in the last-named sum has θ as infimum, since each f_i is summable. Hence by (6.9) the left member of the inequality is $\leqq \theta$. By (11.4) it is $\geqq \theta$, so it is θ, and mid (f_1,f_2,f_3) is summable.

Next let f_1 and f_2 be summable elements of F such that $I(f_1)$ and $I(f_2)$ have a common upper or lower bound. By (6.7) they have a common lower bound b. Let u_1 be in $U[\geqq f_1]$ and u_2 in $U[\geqq f_2]$. Then b is a common lower bound for $I_1(u_1)$ and $I_1(u_2)$, so by (12.4) $u_1 \vee u_2$ is a summable U-element. By the preceding proof, $f_1 \vee f_2 =$ mid $(f_1,u_1 \vee u_2,f_2)$ is summable. Similarly, $f_1 \wedge f_2$ is summable.

Finally, let f_1,f_2,f_3 be any three summable elements such that $I(f_1)$ and $I(f_2)$ have a common lower or upper bound. Then $f_1 \vee f_2$ and $f_1 \wedge f_2$ are summable, and by the first part of this proof mid $(f_1 \wedge f_2,f_1 \vee f_2,f_3)$ is summable. By (2.13), this is the same as mid (f_1,f_2,f_3).

(13.2) **THEOREM**. Let (12.1[σ]) and (12.7) hold. If f_1 and f_2 are summable elements of F such that $I(f_1)$ and $I(f_2)$ have a common bound, then $f_1 \vee f_2$ and $f_1 \wedge f_2$ are summable, and

$$I(f_1 \vee f_2) + I(f_1 \wedge f_2) = I(f_1) + I(f_2) .$$

We already know that $f_1 \vee f_2$ and $f_1 \wedge f_2$ are summable. Let u_i belong to $U[\geqq f_i]$, $I = 1, 2$. Then by (12.8), $u_1 \vee u_2$ is in $U[\geqq f_1 \vee f_2]$ and $u_1 \wedge u_2$ in $U[\geqq f_1 \wedge f_2]$, and

$$I(f_1 \vee f_2) + I(f_1 \wedge f_2) \leqq I_1(u_1) + I_1(u_2) .$$

Since u_i is an arbitrary member of $U[\geqq f_i]$, this implies

$$I(f_1 \vee f_2) + I(f_1 \wedge f_2) \leqq I(f_1) + I(f_2) .$$

The reverse inequality is obtained by considering elements of $L[\leqq f_i]$, $i = 1, 2$, which completes the proof.

As an aid in verifying one of the hypotheses in the two preceding theorems, we establish a lemma.

(13.3) LEMMA. If E is directed by $\geqq$ or by $\leqq$ and f_1 and f_2 are summable, then $I(f_1)$ and $I(f_2)$ have a common upper bound and a common lower bound.

Choose u_i in $U[\geqq f_i]$ and $e_i \ll u_i$, $e = 1, 2$. By hypothesis, E contains an element e satisfying either $e \geqq e_i$ $(e = 1, 2)$ or $e \leqq e_i$ $(i = 1, 2)$. Hence $I_0(e_1)$ and $I_0(e_2)$ have a common upper or lower bound in G; by (6.7) they have a common upper bound. Each pair of consecutive elements of the finite sequence $I(f_1)$, $I_1(u_1)$, $I_0(e_1)$, $I_0(e_2)$, $I_1(u_2)$, $I(f_2)$ has an upper bound in G, so by (6.7) all six have a common upper and lower bound.

§ 14. MONOTONE SEQUENCES OF SUMMABLE ELEMENTS

In order to establish convergence theorems of adequate strength, we apparently need to add something to postulates (12.1). The convergence theorems will rest on theorems concerning monotone sequences, and in this section we establish two such theorems. The first is of less general applicability, but is not entirely without interest.

Our first strengthening of (12.1) is as follows.

(14.1) POSTULATE. G is a group under $+$ with the property that whenever $S_1, S_2, \ldots$ are countably many subsets of G directed by $\leqq$ and having $\wedge S_i = \theta$ for $i = 1, 2, \ldots$, it is possible to choose elements $g_1, g_2, \ldots$ such that g_i is in S_i and the sums $g_1 + \ldots + g_k$ $(k = 1, 2, \ldots)$ have an upper bound.

For example, the reals have this property; but the group of bounded real functions on an infinite domain D lacks the property, as we see if we take each S_i to consist of all the functions which vanish on finite subsets of D and are 1 on the rest of D.

(14.2) COROLLARY. Let (12.1[σ]) and (14.1) hold. Let $S_1, S_2, \ldots$ be countably many subsets of G directed by $\leqq$ and having $\bigwedge S_i = \theta$, and let S consist of all sums of convergent series $\Sigma_i g_i$ with g_i in S_i. Then S is directed by $\leqq$, and $\bigwedge S = \theta$.

By (14.1), there is at least one convergent series, say $\Sigma_i g_i'$. Let S' be the sum of this series. It is easy to verify that S is directed by $\leqq$. For each n, the set of elements $g_1 + \cdots + g_n + \Sigma_{j=n+1}^{\infty} g_j'$ with $g_1, \ldots, g_n$ in $S_1, \ldots, S_n$ respectively is directed by $\leqq$, and by (6.9) its infimum is $\Sigma_{j=n+1}^{\infty} g_j'$. This set is contained in S, so

$$\theta \leqq \bigwedge S \leqq \sum_{j=n+1}^{\infty} g_j' , \quad n = 1, 2, 3, \ldots .$$

By (6.8)

$$\begin{aligned} \bigwedge_n \sum_{j=n+1}^{\infty} g_j' &= g' - \bigvee_n (g' - \{g_1' + \cdots + g_n'\}) \\ &= \theta , \end{aligned}$$

completing the proof.

(14.3) THEOREM. Let (12.1[σ]) and (14.1) hold. Let $f_1, f_2, \ldots$ be an isotone sequence of summable elements of F, and let $f_o = \lim_n f_n$. Then f_o is summable if and only if the $I(f_n)$ have an upper bound; and in that case, $I(f_o) = \lim_{n \to \infty} I(f_n)$.

If f_o is summable the $I(f_n)$ have $I(f_o)$ as upper bound. Suppose then that the $I(f_n)$ have an upper bound; by the Dedekind completeness they have a least upper bound g, and by Remark 2 after (3.1), $g = \lim_{n \to \infty} I(f_n)$. For each n, the set S_n of elements $\{I_1(u' - I_1(l') : u'$ in $U[\geqq f_n]$ and l' in $L[\leqq f_n]\}$ is a subset of G directed by $\leqq$, according to (11.2), and by (6.8) $\bigwedge S_n = \theta$. Let S consist of all convergent sums $\Sigma_n G_n$ with g_j in S_j. For each such sum, by definition of S_n there exists for each n a u_n' in $U[\geqq f_n]$ and an

l_n' in $L[\leqq f_n]$ such that $I_1(u_n') - I_1(l_n') = g_n$. Now define $u_n = \bigvee\{u_1', \ldots, u_n'\}$, $l_n = \bigvee\{l_1', \ldots, l_n'\}$. By (12.4) and (10.5), u_n is a summable U-element and l_n is a summable L-element, and clearly $l_n \leqq f_n \leqq u_n$, $(n = 1, 2, \ldots)$. By induction we prove

$$I_1(u_n) \leqq \sum_{j=1}^{n} g_j + I_1(l_n) ;$$

for when $n = 1$ the two sides are identical, and if the equation holds for $n = k - 1$ then by (12.6) and the definitions of u_n and l_n

$$\begin{aligned} I_1(u_k) - I_1(l_k) &\leqq I_1(u_{k-1}) - I_1(l_{k-1}) \\ &\quad + I_1(u_k') - I_1(l_k') \\ &\leqq (g_1 + g_2 + , \ldots + g_{k-1}) + g_k . \end{aligned}$$

Since $I_1(l_n) \leqq I(f_n) \leqq g$, this implies

$$I_1(u_n) \leqq \sum_{j=1}^{\infty} g_j + g .$$

Now by (10.7) $\bigvee_n u_n$ is a summable U-element, and

$$I_1(\bigvee_n u_n) = \bigvee_n I_1(u_n) \leqq \sum_{j=1}^{\infty} g_j + g .$$

But $\bigvee_n u_n \geqq \bigvee_n f_n = f_0$, so $\bigvee_n u_n$ is in $U[\geqq f_0]$, and so

$$\bar{I}(f_0) \leqq \sum_{j=1}^{\infty} g_j + g .$$

By (14.2) and (6.8),

$$\bar{I}(f_0) \leqq g .$$

On the other hand, since $f_n \leqq f_0$ for all n,

$$I(f_n) \leqq \underline{I}(f_0) , \quad \text{so} \quad g = \lim I(f_n) \leqq \underline{I}(f_0) .$$

Hence f_0 is summable, and $I(f_0) = g = \lim I(f_j)$.

The somewhat restrictive requirement (14.1) can be replaced by the much weaker assumption of normality of G (cf. (5.9)), provided that we strengthen the hypotheses on the sequence $(f_n: n = 1, 2, \ldots)$:

(14.4) **THEOREM.** Let (12.1[σ]) hold, and let G be normal. Let $f_1, f_2, \ldots$ be an isotone sequence of summable elements such that there exists a summable U-element u which satisfies $u \geqq f_n$, $n = 1, 2, \ldots$. Let $f_0 = \lim_{n \to \infty} f_n$. Then f_0 is summable, and $I(f_0) = \lim_{n \to \infty} I(f_n)$.

As before, we define g to be $\lim_{n \to \infty} f_n$. Clearly $\underline{I}(f)$ and $\overline{I}(f)$ are defined, and by (11.5) $\underline{I}(f) \geqq g$.

For each n, the elements $I_1(u'_n) - I_1(l'_n)$ with u'_n in $U[\geqq f_n]$ and l'_n in $L[\leqq f_j]$ form a subset of G directed by $\leqq$, according to (11.2); and by (6.8) their infimum is the zero-element θ, since f_n is summable. Let ϵ be an arbitrary positive number, and let r belong to the class R_G of smoothly rising real functions on G. Because of the continuity of r, we can choose u'_1 in $U[\geqq f_1]$ and l'_1 in $L[\leqq f_1]$ so that

$$r([I_1(u'_1) - I_1(l_1')] + g) < r(g) + \epsilon .$$

Then, step by step, we can choose u'_n in $U[\geqq f_n]$ and l'_n in $L[\leqq f_n]$ so that

$$(*) \qquad r\left(\sum_{j=1}^{n} [I_1(u'_j) - I_1(l'_j)] + g\right) < r(g) + \epsilon$$

for all positive integers n. Without loss of generality we may assume $u'_n \leqq u$ for each n, since we may replace u'_n by $u'_n \wedge u$.

Next we define $u_n = \bigvee\{u'_1, \ldots, u'_n\}$, $l_n = \bigvee\{l'_1, \ldots, l'_n\}$. Then u_n is a U-element by (10.6), and is summable because it is $\leqq u$; and l_n is a summable L-element by the dual of (10.5). Also $l_n \leqq f_n \leqq u_n$ $(n = 1, 2, \ldots)$.

As in the proof of (14.2), by induction we prove

$$I_1(u_n) \leqq \sum_{j=1}^{n} [I_1(u'_n) - I_1(l'_n)] + I_1(l_n) .$$

Since r is isotone and $I_1(l_n) \leqq I(f_n) \leqq g$, this implies

$$r(I_1(u_n)) \leqq r\left(\sum_{j=1}^{n} [I_1(u'_n) - I_1(l'_n)] + g\right).$$

But from this and (*),

$$r(I_1(u_n)) \leqq r(g) + \epsilon .$$

Now by (10.7) $\bigvee_n u_n$ is a U-element. It is summable, being $\leqq u$; and since $\bigvee_n u_n \geqq u_j \geqq f_j$ for each j, $\bigvee_n u_n \geqq \bigvee f_j = f_0$, and it is therefore in $U[\geqq f_0]$. Since r is smoothly rising and consequently is o-continuous,

$$r(I_1(\bigvee_n u_n)) = \bigvee_n r(I_1(u_n))$$
$$\leqq r(g) + \epsilon .$$

But $\bigvee_n u_n$ is in $U[\geqq f_0]$, so $I_1(\bigvee_n u_n) \geqq \overline{I}(f_0)$, and

$$r(\overline{I}(f_0)) \leqq r(g) + \epsilon .$$

Since ϵ is arbitrary, $r(\overline{I}(f)) \leqq r(g)$. Here r is any member of R_G, so by the normality of G (cf. (5.9)) $\overline{I}(f) \leqq g$. With (11.4) and the previously established inequality $\underline{I}(f) \geqq g$, this implies $\overline{I}(f) = \underline{I}(f) = g$, and the proof is complete.

It would seem reasonable to expect that the hypothesis of the existence of a summable $u \geqq f_n$ could be weakened to read "the $I(f_n)$ have an upper bound in G." The impossibility of such a weakening is shown by the following example. F consists of all extended-real functions on $[0,1]$, G of all (finite) real functions on $[0,1]$, E of all continuous functions on $[0,1]$. The partial order $\geqq$ has the usual meaning, and $\gg$ is the same as $\geqq$. I_0 is the identical mapping. We find that u is a U-element if and only if it is lower semi-continuous and bounded below, and dually for L-elements; a U- or L-element is summable if and only if it is finite-valued. Now let the rationals in $[0,1]$ be arranged in a sequence $t_1, t_2, \ldots$, and let f_n be defined to equal i at t_i, $i = 1, \ldots$, and to be 0 elsewhere. Each f_n is a summable L-function, and $I(f_n) = I_1(f_n) = f_n$. The supremum of $I(f_n)$ is the element f, defined by the equation $f(t_i) = i$ $(i = 1, 2, \ldots)$, $f(t) = 0$ elsewhere. But $\bigvee_n f_n$ is not summable. If it were, there would be a summable U-element $u \geqq \bigvee_n f_n$. This u is lower semi-continuous and finite-valued, but is unbounded on every interval. This contradicts a well-known property of semi-continuous functions.

§ 15. DOMINATED CONVERGENCE

The results in the preceding section allow us to establish a generalization of the Lebesgue dominated-convergence theorem.

(15.1) **THEOREM.** Let (12.1[σ]) hold, and let G be normal. Let $(f(\alpha):\alpha$ in $A)$ be a net of elements and f_o an element of A with the following properties.

(i) For each α in A, $f(\alpha)$ is summable.

(ii) There exist summable elements f', f'' of F such that $f' \leqq f(\alpha) \leqq f''$ for all α in A.

(iii) The net $(f(\alpha):\alpha$ in $A)$ is σo-convergent to f_o.

Then the following conclusions are valid.

(iv) f_o is summable.

(v) The net $(I(f(\alpha)):\alpha$ in $A)$ is σo-convergent to $I(f_o)$.

Furthermore, the theorem remains valid if we replace σo by σo* in hypothesis (iii) and conclusion (v).

Let S be the set of summable elements s of F such that $f' \leqq s \leqq f''$. If K is a countable subset of S directed by $\geqq$, by (2.4) there is an isotone sequence $f_1, f_2, \ldots$ of elements of K such that $\bigvee_n f_n = \bigvee K$. By (14.4) (with u any summable U-element $\geqq f''$) we have $\bigvee K$ in S, and $I(\bigvee K) = \bigvee_n(I(f_n)$; and it is easy to verify that $\bigvee_n I(f_n) = \bigvee I(K)$. Similarly we show that if K is a countable subset of S directed by $\leqq$, then $\bigwedge K$ is in S and $I(\bigwedge K) = \bigwedge I(K)$. So S is Dedekind σ-closed and I is σ-smoothly rising on S. By (13.1) S is a lattice; by (4.3) it is σ-complete. By (5.6) the mapping I is σo-continuous, and by (5.2) it is σo*-continuous, on S. This establishes the theorem.

The following corollary is in a form frequently convenient.

(15.2) COROLLARY. Let (12.1[σ]) hold, and let G be normal. Let $f_1, f_2, \ldots$ be a sequence of summable elements of F all between two summable elements f', f'' and o-convergent to an element f_o of F. Then f_o is summable, and o-lim $I(f_n)$ exists, and is equal to $I(f_o)$.

We have assumed that G is normal, that is that there are enough functions r in R_G to determine order in G. If there are enough functions in R_G to determine o-convergence in G, we can strengthen the conclusion of (15.1) somewhat, as follows.

(15.3) THEOREM. Let (8.1[σ]) hold and let G be normal. Assume that whenever g_o is in G and $(g(\alpha):\alpha$ in $A)$

is an eventually dominated net of elements of G such that $\lim r(g(\alpha)) = r(g_0)$ for all r in R_G, it is also true that $o\text{-}\lim g(\alpha) = g_0$. Let $(f(\alpha):\alpha$ in $A)$ be a net of summable elements of F and f_0 an element of F such that

(i) there exist summable elements f', f'' of F such that $f' \leqq f(\alpha) \leqq f''$ for all α in A;

(ii) $\sigma o^*\text{-}\lim f(\alpha) = f_0$.

Then f_0 is summable, and $(I(f(\alpha)):\alpha$ in $A)$ is o-convergent, and $o\text{-}\lim I(f(\alpha)) = I(f_0)$.

By (3.10), there is a sequence $(f(\alpha_n):\ n = 1, 2, \ldots)$ which is o-convergent to f_0. So by (15.2) f_0 is summable. Now let r belong to R_G; we shall establish

(*) $$\lim r(I(f(\alpha))) = r(I(f_0)) \ .$$

If this were false, for some positive ϵ the subset B consisting of those points β of A at which

(**) $$|\cdot r(I(f(\beta))) - r(I(f_0)) \ | \geqq \epsilon$$

would be cofinal in A. Taking $\alpha(\beta) = \beta$, the system $(f(\alpha(\beta)):\beta$ in B) is a subnet of $(f(\alpha):\alpha$ in A). By (3.10), there exists a sequence $\beta_1, \beta_2, \ldots$ of points of B such that $o\text{-}\lim f(\alpha(\beta_n)) = f_0$. By (15.2), and the continuity of r,

$$\lim_{n \to \infty} r(I(f(\alpha(\beta_n)))) = r(I(f_0)) \ .$$

But this contradicts (**), so (*) holds. By hypothesis, this implies $o\text{-}\lim I(f(\alpha)) = I(f_0)$.

We showed in (11.7) that the function I on the class of summable elements is an extension of the function I_1 on the class of summable U- or L-elements. However, we have not yet investigated whether I is an extension of the elementary mapping I_0 from which it was constructed. As a matter of fact, without further hypotheses this may fail to be the case. We accordingly state a new condition which I_0 may or may not satisfy.

(15.4) POSTULATE. To the element e_0 of E correspond three sequences e'_i, e''_i, e'''_i $(i = 1, 2, \ldots)$ of elements of E with the following properties.

(i) There exist elements f, g of F, each of which is either summable or in E, such that $f \leqq e_i' \ll e_i'' \ll e_i''' \leqq g$ ($i = 1, 2, \ldots$).

(ii) $\text{o-lim } e_i' = \text{o-lim } e_i''' = e_0$.

(iii) $\text{o-lim } I_0(e_i'') = I_0(e_0)$.

With this we establish the equality of I and I_0 on E.

(15.5) THEOREM. If (8.1[σ]) holds, and G is normal, and condition (15.4) is satisfied for an e_0 in E, then e_0 is summable, and $I(e_0) = I_0(e_0)$.

By (9.7), for each i there are summable U-elements u', u_i'' such that $e_i' \ll u_i' \ll e_i'' \ll u_i'' \ll e_i'''$. By (9.6), $I_1(u_i') \leqq I_0(e_i'') \leqq I_1(u_i'')$. By (11.7), this implies

(*) $$I(u_i') \leqq I_0(e_i'') \leqq I(u_i'') \ .$$

Corresponding to the f, g of (15.4), there exist summable elements f_0, g_0 such that $f_0 \leqq f$, $g \leqq g_0$; for if g is summable we can choose $g_0 = g$, and if g is in E there is a summable U-element $g_0 \gg g$, by (9.7). Hence $f_0 \leqq e_i' \ll u_i' \ll u_i'' \ll e_i''' \leqq g_0$. From (15.4ii), $\text{o-lim } u_i' = \text{o-lim } u_i'' = e_0$. So by (15.1) $\text{o-lim } I(u_i')$ and $\text{o-lim } I(u_i'')$ exist, and

$$\text{o-lim } I(u_i') = \text{o-lim } I(u_i'') = I(e_0) \ .$$

But by (*), $I_0(e_i'')$ is o-convergent to this same limit;

$$\text{o-lim } I_0(e_i'') = I(e_0) \ .$$

Now from (15.4iii) we obtain $I(e_0) = I_0(e_0)$, concluding the proof.

(15.6) THEOREM. If for an e_0 in E it is true that $\bigwedge\{I_0(e') : e' \text{ in } E \text{ and } e' \gg e_0\} = \bigvee\{I_0(e) : e \text{ in } E \text{ and } e \ll e_0\}$, then e_0 is summable, and $I(e_0) = I_0(e_0)$. In particular, if $\gg$ is reflexive, every e in E is summable, and $I(e) = I_0(e)$.

By (8.2d) the set $S' \equiv \{e' : e' \text{ in } E \text{ and } e' \gg e_0\}$ is directed by $\ll$, so $\bigwedge I_0(S')$ exists and is $\geqq I_0(e_0)$. If e' is in S', by (9.7) there is a U-element u' such that $e_0 \ll u' \ll e'$. Then $I_0(e_0) \leqq I_1(u') \leqq I_0(e')$, so $\bar{I}(e_0) \leqq I_0(e')$, and because e' is an arbitrary number of S', $\bar{I}(e_0) \leqq \bigwedge I_0(S')$. Dually, if S is the set $\{e: e$ in E and $e \ll e_0\}$, then $\underline{I}(e_0) \geqq \bigvee I_0(S) = \bigwedge I_0(S')$, and also $\bigvee I_0(S) \leqq I_0(e_0)$. These inequalities establish the conclusion. If $\gg$ is reflexive, both S and S' contain e_0, so $\bigwedge I_0(S') = I_0(e_0) = \bigvee I_0(S)$, and as just proved this establishes that $I(e_0)$ exists and is equal to $I_0(e_0)$.

We now return to sequences of functions, and establish some easy theorems, the last of which is a generalization of Fatou's lemma.

(15.7) **THEOREM**. Let (12.1[σ]) hold. Let G be normal, or let (14.1) hold. If S is a countable set of summable elements directed by $\leqq$ and S has a summable lower bound f', then $\bigwedge S$ is summable, and $I(\bigwedge S) = \bigwedge I(S)$.

By (2.4), there is a sequence $f_1 \geqq f_2 \geqq f_3 \geqq \cdots$ of summable elements of S such that $\bigwedge_n f_n = \bigwedge S$. Since $f_1 \geqq f_j \geqq f'$, $(j = 1, 2, \ldots)$, the conclusion holds by (15.2).

(15.8) **THEOREM**. Let (12.1[σ]) hold. Let G be normal, or let (14.1) hold. If S is a countable set of summable elements of F, all $\geqq$ a summable element f', then $\bigwedge S$ is summable.

Let S consist of all infima of finite subsets of S. By (13.1) all elements of S^* are summable. S^* is directed by $\leqq$, and by (1.2) we have $\bigwedge S = \bigwedge S^*$, which is summable by (15.7).

(15.9) **THEOREM**. Let (12.1[σ]) and (14.1) be satisfied. Let $f, f_1, f_2, f_3, \ldots$ be summable elements of F such that $f_n \geqq f$ $(n = 1, 2, \ldots)$. Assume that there exists an element g' of G such that $I(f_n) \leqq g'$ for infinitely many n (or, more particularly, that $\liminf I(f_n)$ exists). Then $\liminf f_n$ is summable, and $I(\liminf f_n) \leqq g'$; in particular, if $\liminf I(f_n)$ exists, then $I(\liminf f_n) \leqq \liminf I(f_n)$.

For each positive integer m, define $f_m' = \bigwedge\{f_m, f_{m+1}, \ldots\}$. This is summable by (15.8), and $f \leqq f_m' \leqq f_i$ if $1 \leqq m \leqq i$. Hence $I(f_m') \leqq I(f_i)$ if $i \geqq m$, and for some $i \geqq m$ we have $I(f_i) \leqq g'$, and moreover $I(f_m') \leqq \liminf I(f_j)$ if this last exists. Since $f_1' \leqq f_2' \leqq \cdots$, by (14.3) $\bigvee_m f_m'$ is summable, and $I(\bigvee_m f_m') = \bigvee_m I(f_m') \leqq g'$, and $I(\bigvee_m f_m') \leqq \liminf_n I(f_n)$ if this exists. But $\bigvee_m f_m' = \liminf f_j$, so the proof is complete.

§ 16. STRUCTURE OF THE SET OF SUMMABLE ELEMENTS

The information about summable elements obtained in the preceding sections can be combined into a theorem on the structure of the set of summable elements of F. Recalling the discussion following (6.7), we note that each element g of G determines a subset G_g of G consisting of all elements g' such that g and g' have a common upper or lower bound; two such classes are either disjoint or identical, and each is in fact a coset relative to the invariant subgroup G_θ. Correspondingly, we adopt a notation.

(16.1) DEFINITION. Let (8.1[σ]) hold. F_{sum} is defined to be the set of all summable elements of F. For each g in G, F_g is the set of all summable elements f such that $I(f)$ is in G_g.

(16.2) THEOREM. Let (8.1[σ]) hold and let G be normal. Then F_{sum} is the union of disjoint classes F_g with the following properties.

(i) If f, f' belong to distinct classes F_g, $F_{g'}$, the pair $\{f, f'\}$ has neither an upper bound nor a lower bound in F_{sum}.

(ii) F_{sum} is Dedekind σ-complete.

(iii) Each class F_g is a conditionally σ-complete lattice.

(iv) If (15.4) holds, F_{sum} contains E, and for all e in E the equation $I(e) = I_o(e)$ holds.

(v) I is a σo-continuous and σo*-continuous map of F_{sum} into G.

(vi) If E is directed by $\geqq$ or by $\leqq$, there is only one non-empty class F_g, and this coincides with F_{sum}.

If f and f' have a common upper bound f'' in F_{sum}, then $I(f)$ and $I(f')$ have the upper bound $g = I(f'')$ in G, and f and f' are in the same set F_g. This establishes (i). Conclusion (ii) follows from (15.7) and its dual. This also implies the Dedekind-completeness of each F_g, since with the notation of (15.7), for each f in S, I(f) and $I(\bigwedge S)$ have the common lower bound I(f'). Each F_g is a lattice, by (13.1); by (2.3) and the preceding sentence, it is conditionally σ-complete. Conclusion (iv) was established in (15.5). Since σo-convergence of $(f(\alpha):\alpha$ in A) in F_{sum} (as distinguished from σo-convergence in F) implies that $f(\alpha)$ is eventually between summable elements, (v) follows from (15.1). Finally, (vi) follows from (13.1), (13.3) and (i).

§ 17. MEASURABLE ELEMENTS

Like M. H. Stone, we define measurability with the help of the concept of "middle."

(17.1) DEFINITION. Let (8.1[σ]) hold. An element f of F is measurable if for each pair of summable elements f',f" with $f' \leqq f''$ it is true that mid (f',f",f) is summable.

(17.2) COROLLARY. If f is measurable and there are summable elements f',f" such that $f' \leqq f \leqq f''$, then f is summable.

For then f = mid (f',f",f), which is summable by hypothesis.

(17.3) COROLLARY. If f is summable, it is measurable.

This follows at once from (13.1).

The next corollary is a special case of the following theorem, but it is easily proved and covers at least some interesting cases.

(17.4) COROLLARY. If (12.1[σ]) holds, and to each summable element f corresponds a summable U-element $u \leqq f$, then every U-element is measurable.

Let u be a U-element, and let f' and $f'' \geqq f'$ be summable. There exists a summable $u'' \geqq f''$, and by hypothesis there exists a summable $u' \leqq f'$. By (10.8), mid (u',u",u) is summable. By (13.1), mid (f',f",mid (u',u",u)) is summable. But mid (f',f",mid (u',u",u)) = $f' \vee \{f'' \wedge [u'' \wedge (u' \vee u)]\} = f' \vee \{f'' \wedge (u' \vee u)\} = (f' \vee f'') \wedge (f' \vee [u' \vee u]) = (f' \vee f'') \wedge (f' \vee u) = f' \vee (f'' \wedge u)$ = mid (f',f",u).

(17.5) THEOREM. If (12.1[σ]) holds, every U-element is measurable, and so is every L-element.

To be specific, we consider a U-element f_3. Let f' and f'' be any two summable elements with $f'' \geqq f'$. Let f_1 be in $L[\leqq f']$ and f_2 in $U[\geqq f'']$. The computation in (17.4), with obvious notational changes, shows that mid (f', f'', f_3) = mid $(f', f'', \text{mid } (f_1, f_2, f_3))$. So by (13.1) and (17.3) it is sufficient to prove that mid (f_1, f_2, f_3) is summable.

Let $S^{(i)}$ $(i = 1, 2, 3)$ be three [countable] subsets of E, the first directed by $\ll$ and the other two by $\gg$, such that $\bigwedge S^{(1)} = f_1$, $\bigvee S^{(2)} = f_2$, $\bigvee S^{(3)} = f_3$. For each pair $e_1^{(i)}$ and $e_2^{(i)}$ of elements of $S^{(i)}$ such that $e_1^{(i)} \ll e_2^{(i)}$, we choose exactly one U-element and exactly one L-element between $e_1^{(i)}$ and $e_2^{(i)}$; this is possible by (9.7). The chosen U-elements form the set $S_U^{(i)}$: the chosen L-elements form the set $S_L^{(i)}$. If u_1, u_2 are in $S_U^{(1)}$, $S_U^{(2)}$ respectively, $u_1 \vee u_2$ and $u_1 \wedge u_2$ are summable U-elements by (12.4) and (10.5). By (10.8) and (2.13), mid (u_1, u_2, u_3) is a summable U-element whenever u_i is in $S_U^{(i)}$ $(i = 1, 2, 3)$. Likewise, whenever l_i is in $S_L^{(i)}$ $(i = 1, 2, 3)$, both mid (l_1, l_2, l_3) and mid (f_1, l_2, l_3) are summable L-elements.

Let λ_1, l_2, l_3 be arbitrary members of $S_L^{(1)}$, $S_L^{(2)}$, $S_L^{(3)}$ respectively. By the manner of definition of these sets, there exist elements u_i in $S_U^{(i)}$ $(i = 1, 2, 3)$ such that $u_1 \leqq \lambda_1, u_2 \geqq l_2, u_3 \geqq l_3$ and there exist elements λ_2, λ_3 with λ_i in $S_L^{(i)}$ such that $\lambda_i \geqq u_i$. Then

$$\text{mid } (f_1, l_2, l_3) \leqq \text{mid } (u_1, u_2, u_3) \leqq \text{mid } (\lambda_1, \lambda_2, \lambda_3) .$$

Since $f_1 \leqq f_2 \geqq l_2$, it follows that mid $(f_1, l_2, l_3) \leqq f_2$, so the elements $I_1(f_2), I_1(f_1), I_1(\text{mid } [f_1, l_2, l_3])$ have a common bound in G. Also, mid $(f_1, l_2, l_3) \leqq$ mid (f_1, f_2, f_3) and $f_1 \leqq$ mid (f_1, f_2, f_3), so $[\text{mid } (f_1, l_2, l_3)] \vee f_1$ is in $L[\leqq \text{mid } (f_1, f_2, f_3)]$, and

$$I_1[\text{mid } (f_1, l_2, l_3) \vee f_1] \leqq \underline{I}[\text{mid } (f_1, f_2, f_3)] .$$

Since $\lambda_1 \geqq f_1$, by the distributive law

$$\begin{aligned}[\text{mid } (f_1, \lambda_2, \lambda_3)] \vee \lambda_1 &= [(f_1 \vee \lambda_2) \wedge (f_1 \vee \lambda_3) \wedge (\lambda_2 \vee \lambda_3)] \vee \lambda_1 \\ &= (\lambda_1 \vee \lambda_2) \wedge (\lambda_1 \vee \lambda_3) \wedge (\lambda_1 \vee \lambda_2 \vee \lambda_3) \\ &\geqq \text{mid } (\lambda_1, \lambda_2, \lambda_3) .\end{aligned}$$

Hence, using (12.6),

$$\begin{aligned} I_1[\text{mid }(u_1,u_2u_3)] &\leqq I_1[\text{mid }(\lambda_1,\lambda_2,\lambda_3)] \\ &\leqq I_1[\text{mid }(f_1,\lambda_2,\lambda_3)\vee\lambda_1] \\ &\leqq I_1[\text{mid }(f_1,\lambda_2,\lambda_3)\vee f_1] + I_1(\lambda_1) - I_1(f_1) \\ &\leqq \underline{I}\ [\text{mid }(f_1,f_2,f_3)] + I_1(\lambda_1) - I_1(f_1)\ . \end{aligned}$$

The [countable] set of elements $\{\text{mid }(u_1,u_2,u_3)\colon u_2$ in $S_U^{(2)}$ and u_3 in $S_U^{(3)}\}$ is directed by $\geqq$ and has $\text{mid }(u_1,f_2,f_3)$ as supremum. Hence by (10.6)

$$I_1(\text{mid }(u_1,f_2,f_3)) \leqq \underline{I}[\text{mid }(f_1,f_2,f_3)] + [I_1(\lambda_1) - I_1(f_1)]\ .$$

But $\text{mid }(u_1,f_2,f_3)$ is in $U[\geqq \text{mid }(f_1,f_2,f_3)]$, hence

$$\overline{I}(\text{mid }(f_1,f_2,f_3)) \leqq \underline{I}[\text{mid }(f_1,f_2,f_3)] + [I_1(\lambda_1) - I_1(f_1)]\ .$$

The set of values of the quantity in brackets corresponding to all λ_1 in $S_L^{(j)}$ has infimum θ, so by (6.9)

$$\overline{I}[\text{mid }(f_1,f_2,f_3)] \leqq \underline{I}[\text{mid }(f_1,f_2,f_3)]\ .$$

This proves that $I[\text{mid }(f_1,f_2,f_3)]$ exists, and the proof is complete.

(17.6) **THEOREM.** Let (12.1[σ] hold. If $(f(\alpha)\colon\alpha$ in $A)$ is a net of measurable elements and is σo- or σo*-convergent to a limit f_0 in F, then f_0 is measurable.

Let f' and $f'' \geqq f'$ be summable elements of F. By (5.3) and (5.2), $\text{mid }(f(\alpha),f',f'')$ is σo- or σo*-convergent to $\text{mid }(f_0,f',f'')$. By (15.1) this is summable. So f_0 is measurable.

(17.7) **THEOREM.** Let (8.1[σ]) hold. The set F_{meas} of measurable elements f is a σ-complete lattice, and contains all U-elements, all L-elements and all summable elements.

Let f_1 and f_2 be measurable, and let f' and $f'' \geqq f'$ be summable. Then by (2.14) $\text{mid }(f',f'',f_1\vee f_2) = [\text{mid }(f',f'',f_1)] \vee [\text{mid }(f',f'',f_2)]$, which is summable by (13.1). So $f_1\vee f_2$ is measurable. Similarly, $f_1\wedge f_2$ is measurable, so F_{meas} is a lattice. By

(17.6) and (4.3), F_{meas} is σ-complete. The other statements were established in (17.3) and (17.5).

(17.8) COROLLARY. If $f_1, f_2, f_3, \ldots$ are measurable elements of F, so are $\bigvee\{f_i : i = 1, 2, \ldots\}$, $\bigwedge\{f_i : i = 1, 2, \ldots\}$, o-lim sup f_i and o-lim inf f_i.

The first two statements follow from (17.7). In particular, $f'_n = \bigvee\{f_j : j = n, n+1, \ldots\}$ is measurable, hence so is o-lim inf $f_i \doteq \bigwedge\{f'_n : n = 1, 2, \ldots\}$. Dually, o-lim inf f_i is measurable.

CHAPTER IV

ALGEBRAIC OPERATIONS

§ 18. INTERCHANGE OF ORDER OF OPERATIONS

The theorems to be established in this chapter are of the type "the integral of a sum is the sum of the integrals." In each case, we have an operation Φ that can be performed on certain summable elements f, and an operation Γ that can be performed on certain elements of G. We are interested in knowing when these are permutable, so that $\Gamma(I(f)) = I(\Phi(f))$. Our procedure is to postulate this permutability for elements of E and then extend it to the summable elements of F. However, to gain generality enough for our uses it is desirable to consider mappings of one system on another, not merely mappings of a system into itself. Also, in discussing the mapping Φ we shall start with an elementary mapping Φ_0 and extend it in the same way as we extended I_0. But the classes of summable elements arrived at by starting with Φ_0 are not necessarily the same as those reached by starting with I_0. To distinguish these classes, we shall use "Φ" or "I" in their name; thus the summable elements defined by extension of I_0 are now to be called "I-summable elements."

Our principal theorem on interchange of order of operations is the following.

(18.1) THEOREM. Let $\{F, \gg, E, G, I_0\}$ and $\{F', \gg', E', G', I_0'\}$ be two systems satisfying (8.1[σ]). Assume further

(a) Φ_0 is an isotone function on a domain which is contained in F' and contains E', and whose image is in F; and for any two members e_1', e_2' of E', $\Phi_0(e_1')$ is in E, and if $e_1' \gg' e_2'$ then $\Phi_0(e_1') \gg \Phi_0(e_2')$; and with the partial ordering $\gg'$, Φ_0 satisfies (8.1e).

(b) Γ is a smoothly rising function whose domain is a Dedekind-closed subset D_Γ of G' and whose range is in G; and for each e' in E', $I'_0(e')$ is in D_Γ.

(c) For each e' in E', $\Gamma(I_0'(e')) = I_0(\Phi_0(e'))$.

Let f' be an I'-summable member of F', and let $E'_{f'}$ consist of those elements e' of E' such that $I'(f')$ and $I'_0(e')$ have a common upper or lower bound in G'.

Then the following conclusions are valid.

(i) The system $\{F', \gg', E'_{f'}, F, \Phi_0\}$ with Φ_0 restricted to $E'_{f'}$ satisfies (8.1[σ]), and therefore by the processes of Chapter II can be used to define extended operations $\Phi_1, \bar{\Phi}, \underline{\Phi}$ and Φ just as $I_1, \bar{I}, \underline{I}$ and I were defined.

(ii) $I'(f')$ is in the domain D_Γ.

(iii) $\bar{\Phi}(f')$ and $\underline{\Phi}(f')$ are both I-summable, and $I(\bar{\Phi}(f')) = I(\underline{\Phi}(f')) = \Gamma(I'(f'))$.

(iv) If Φ_0 is smoothly rising on its domain and f' is in the domain of Φ_0, $\Phi_0(f')$ is also I-summable, and $I(\Phi_0(f')) = \Gamma(I'(f'))$.

To establish (i) we need merely verify the postulates, which is easy. Suppose now that u' is an I'-summable U-element of F'. There is a [countable] subset S' of E' directed by $\gg'$ and having $\bigvee S' = u'$. Hence the set $I_0'(S')$ is directed by $\geqq$, and since u' is I'-summable this set has an upper bound. It therefore has a supremum $\bigvee I_0'(S')$, which is in D_Γ because D_Γ is Dedekind-closed. Also, since Γ is smoothly rising,

$$\Gamma(I_1'(u')) = \Gamma(\bigvee I_0(S')) = \bigvee \Gamma(I_0'(S')) .$$

If f' is I'-summable, the set $U[\geqq f']$ is a subset of F' directed by $\leqq$. So $I_1'(U[\geqq f'])$ is a subset of D_Γ directed by $\leqq$; its infimum is $I'(f')$, which is in D_Γ because this is Dedekind-closed. So (ii) is proved.

Again let u' and S' be as in the preceding paragraph. For each e' in S', $\Gamma(I_0'(e')) = I_0(\Phi_0(e'))$. The set $\Phi_0(S') = \{\Phi_0(e'): e' \text{ in } S'\}$ is in E and is directed by $\gg$, according to hypothesis (a). So $\bigvee \Phi_0(S')$ is a U-element of F. By definition, $\Phi_1(u') = \bigvee \Phi_0(S')$. For each e' in S', $I_0(\Phi_0(e')) = \Gamma(I_0'(e')) \leqq$

$\bigvee\Gamma(I_0'(S')) = \Gamma(\bigvee I_0'(S')) = \Gamma(I'(u'))$, so $\bigvee\Phi_0(S')$ is an I-summable U-element. Also, $I_1(\Phi_1(u')) = \bigvee I_0(\Phi_0(S')) = \bigvee\{\Gamma(I_0'(e')): e' \text{ in } S'\} = \Gamma(I_1'(u'_1))$. The dual holds for I'-summable L-elements of F'.

Now let f' be an I'-summable element of F'. By definition, $\bar{\Phi}(f')$ is the infimum of $\Phi_1(u')$ for all Φ-summable elements $u' \geqq f'$; the requirement of Φ-summability may be ignored, because F has a greatest element and all U-elements in F' are Φ-summable. We obtain a subset by using only I'-summable U-elements u'; hence, if we let U'[f'] mean the set of all I'-summable U-elements u' of F' which satisfy $u' \geqq f'$, we have

$$\bar{\Phi}(f') \leqq \bigwedge\{\Phi_1(u'): u' \text{ in } U'[f']\} .$$

In particular, there are I-summable U-elements $\Phi_1(u')$ such that $\Phi_1(u') \geqq \bar{\Phi}(f')$, and similarly there are I-summable L-elements below $\underline{\Phi}(f')$, so $\bar{I}$ and $\underline{I}$ are defined for both of these, and

$$\begin{aligned}\bar{I}(\bar{\Phi}(f')) &\leqq \bigwedge\{\bar{I}(\Phi_1(u')): u' \text{ in } U'[f']\}\\ &= \bigwedge\{I_1(\Phi_1(u')): u' \text{ in } U'[f']\}\\ &= \bigwedge\{\Gamma(I_1'(u')): u' \text{ in } U'[f']\} .\end{aligned}$$

But the set $\{I_1'(u'): u' \text{ in } U'[f']\}$ is directed by $\leqq$ and has an infimum which is $I'(f')$; and each member of the set is already known to be in D_Γ, which is Dedekind closed. So the infimum $I'(f')$ is also in D_Γ, and $\bar{I}(\bar{\Phi}(f')) \leqq \Gamma(I'(f'))$. Dually, $\underline{I}(\underline{\Phi}(f')) \geqq \Gamma(I'(f'))$. It follows at once that $I(\bar{\Phi}(f))$ and $I(\underline{\Phi}(f'))$ both exist and are equal to $\Gamma(I'(f'))$, and (iii) is established.

Finally, suppose that Φ_0 is smoothly rising and that f' is I'-summable and is in the domain of Φ_0. Let u' be any I'-summable U-element such that $u' \geqq f'$, and let S' be a [countable] subset of E' directed by $\gg$ and having $\bigvee S' = u'$. Because Φ_0 is smoothly rising, and the set consisting of f' alone is directed by $\leqq$, we have $\Phi_0(f') \leqq \bigvee\Phi_0(S') = \Phi_1(u')$. The discussion in the preceding paragraph remains valid with $\Phi_0(f')$ in place of $\bar{\Phi}(f')$, and shows that $\bar{I}(\Phi_0(f')) \leqq \Gamma(I'(f'))$ and $\underline{I}(\Phi_0(f')) \geqq \Gamma(I'(f'))$. Hence $\Phi_0(f)$ is I-summable, and $I(\Phi_0(f) = \Gamma(I'(f'))$. This completes the proof.

§ 19. ADDITION, SCALAR MULTIPLICATION AND BINARY MULTIPLICATION

In this section we make the simplest applications of Theorem (18.1). These all apply to situations in which stronger postulates than

(12.1[σ]) are satisfied; these are as follows.

(19.1[σ]) POSTULATE.

(a),(b) Same as (12.1(a),(b)) and as (8.1a,b).

(c) Same as (12.1c): G is a Dedekind complete partially ordered group under a commutative operation +.

(d) Same as (12.1d) and as (8.1d).

(e) See at bottom.

(f),(g) Same as (12.1f,g) and as (8.1f,g).

(h) There exists a commutative additive group F_+ embedded in F which contains E, and is such that (i) whenever e_1 and e_2 are in E, so is $e_1 + e_2$, and $I_o(e_1 + e_2) = I_o(e_1) + I_o(e_2)$; and (ii) if e_1, e_1', e_2, e_2' are in E and $e_1 \ll e_1'$ and $e_2 \ll e_2'$, then $e_1 + e_2 \ll e_1' + e_2'$. The identity element of F_+ will be denoted by θ_F.

(e) For each [countable] subset S of E directed by $\ll$ and such that $\bigwedge S \leqq \theta_F$ and $I_o(S)$ is bounded below, $\bigwedge I_o(S) \leqq \theta$.

(19.2) LEMMA. If postulates (19.1[σ]) hold, so do (12.1[σ]).

In view of (19.1[σ]h), (19.1[σ]e) implies (8.1[σ]e), which is the same as (12.1[σ]e). We now show that (19.1[σ]) also implies (12.1[σ]h).

With the notation of (12.1[σ]h) let $s = (e_1' - e_1) + (e_2' - e_2) + (e_3' - e_3)$. Then $e_i' \leqq e_i + s$, $i = 1, 2, 3$. From $e_4 \geqq \text{mid}\,(e_1, e_2, e_3)$ we obtain $e_4 \geqq e_1 \wedge e_2$, whence by (6.3) $e_4 + s \geqq (e_1 + s) \wedge (e_2 + s)$. Similarly $e_4 + s \geqq (e_2 + s) \wedge (e_3 + s)$ and $e_4 + s \geqq (e_3 + s) \wedge (e_1 + s)$, whence

$$\begin{aligned} e_4 + s &\geqq \text{mid}\,(e_1 + s,\ e_2 + s,\ e_3 + s) \\ &\geqq \text{mid}\,(e_1', e_2', e_3') \\ &\geqq e_5\ . \end{aligned}$$

Therefore

$$I_o(e_5) - I_o(e_4) \leqq I_o(s) = \sum_{i=1}^{3} [I_o(e_i') - I_o(e_i)]\ ,$$

and (12.1[σ]h) is satisfied.

(19.3) **THEOREM**. Let (19.1[σ]) be satisfied. Let f_1 and f_2 be summable elements of F which are also in F_+. Then $-f_1$ and $f_1 + f_2$ are also summable, and

$$I(f_1 + f_2) = I(f_1) + I(f_2) ,$$

$$I(-f_1) = -I(f_1) .$$

Let $F' = F \times F$, $F_+' = F_+ \times F_+$, $G' = G \times G$, $E' = E \times E$. If (e_1,e_2) is in E', define $I_0'(e_1,e_2)$ to be $(I_0(e_1), I_0(e_2))$. Let $(e_1', e_2') \gg' (e_1,e_2)$ mean that $e_1' \gg e_1$ and $e_2' \gg e_2$, and define $\geqq$ analogously in F' and in G'. There is no difficulty in showing that the new system satisfies (19.1[σ]). For each element $f' = (f_1,f_2)$ in F_+', define $\Phi_0(f') = f_1 + f_2$; for each element $g' = (g_1,g_2)$ in G', define $\Gamma(g') = g_1 + g_2$. Then the hypotheses of (18.1[σ]) are all satisfied. By conclusion (iv) of that theorem, whenever f_1 and f_2 are summable and in F_+, $f_1 + f_2$ is summable, and $I(f_1 + f_2) = I(f_1) + I(f_2)$.

To establish the other part of the theorem, we let $F' = F$, $f_+' = F_+'$, $G' = G$, $E' = E$, $I_0' = I_0$, but $\gg'$ and $\geqq'$ are defined to be $\ll$ and $\leqq$ respectively. For each f' in F_+', let $\Phi_0(f') = -f'$; for each g' in G', let $\Gamma(g') = -g'$. The hypotheses of Theorem (18.1[σ]) are satisfied, and by its conclusion (iv) we find that when f_1 is summable and in F_+, $-f_1$ is summable, and $I(-f_1) = -I(f_1)$.

(19.4) **THEOREM**. Let (19.1[σ]) be satisfied, and also the following hypotheses.

(a) There is a scalar multiplication defined in F_+, with which F_+ is a linear system embedded in F.

(b) G is a partially ordered linear system.

(c) Whenever e is in E and c is real, ce is in E, and $I_0(ce) = cI_0(e)$.

(d) Whenever e_1 and e_2 are in E and $e_1 \ll e_2$, then for all $c > 0$ the inequality $ce_1 \ll ce_2$ holds, while for all $c < 0$ we have $ce_1 \gg ce_2$.

Then for every summable member f of F_+ and every real number c, cf is also summable, and $I(cf) = cI(f)$.

First, consider the case $c > 0$. The new system F', etc., is the same as the old. For all f in F_+', we define $\Phi_0(f') = cf'$; for all g' in G', we define $\Gamma(g') = cg'$. The hypotheses of (18.1[σ]) are satisfied, so the present conclusion holds when $c > 0$. If f is summable and in F_+, and $c < 0$, then $-f$ is summable and in F_+ by (19.3), and $|c|(-f)$ is summable by the proof just completed, and also $I(cf) = I(|c|(-f)) = |c|I(-f) = -|c|I(f) = cI(f)$. So the conclusion holds when $c < 0$. For $c = 0$, we write $0f = f - f$ and obtain the conclusion from (19.3).

The next theorem is a trifle more tedious to establish, because binary multiplication is assumed isotone only for non-negative factors.

(19.5) THEOREM. Let (19.1[σ]) be satisfied, and also the following hypotheses.

(a) F_+ is a conditionally [σ-]complete lattice embedded in F.

(b) F_+ is a ring embedded in F; the identity-element of F_+ will be called θ_F.

(c) G is directed by $\geqq$.

(d) G is a group with binary multiplication, and the domain of the binary multiplication is Dedekind closed in $G \times G$; the identity-element of G will be called θ_G.

(e) Whenever e_1 and e_2 are in E, so is e_1e_2, and $I_0(e_1e_2) = I_0(e_1)I_0(e_2)$.

(f) Whenever e_1, e_1', e_2, e_2' are members of E such that $e_1' \gg e_1 \gg \theta_F$ and $e_2' \gg e_2 \gg \theta_F$ the inequality $e_1'e_2' \gg e_1e_2$ is satisfied.

Then for every pair f_1, f_2 of summable elements of F_+, f_1f_2 is summable, and the pair $(I(f_1), I(f_2))$ is in the domain of the binary multiplication in G, and $I(f_1f_2) = I(f_1)I(f_2)$.

We define F', F_+', G', $\gg'$, $\geqq'$, as in the first paragraph of proof of (19.3). But E' shall consist of all pairs (e_1, e_2) with e_i in E and $e_i \gg \theta_F$ $(i = 1, 2)$. For (e_1, e_2) in E' we define $I_0'(e_1, e_2) = (I_0(e_1), I_0(e_2))$.

If $u' = (u_1, u_2)$ is an I'-summable U-element of F', there is a [countable] subset S' of E' directed by $\gg$ and having $\bigvee S' = u'$, and $\bigvee I_0'(S') \leqq I_1'(u')$. Let S_1 be the set of all first members e_1 of the pairs (e_1, e_2) which constitute S', and let g_1 be the first member of the pair (g_1, g_2) which is $I_1'(u')$. Then S_1 is [countable and] directed by $\gg$, and $\bigvee S_1 = g_1$. So u_1 is an I-summable member of F, and so likewise is u_2, and $I_1'(u') = (I_1(u_1), I_1(u_2))$, and also, $u_i \gg \theta_F$ $(i = 1, 2)$.

Conversely, let u_1 and u_2 be summable U-elements of F such the $u_i \gg \theta_F$, i = 1, 2. Form [countable] subsets S_1, S_2 of E consisting of elements $\gg \theta_F$, directed by $\gg$ and having $\bigvee S_i = u_i$, t = 1, 2; this is possible by (10.4). Then the set S' of all pairs (e_1, e_2) with e_1 in S_1 and e_2 in S_2 is a subset of E', is directed by $\gg'$ and has $\bigvee S' = (u_1, u_2)$. Therefore the I'-summable U-elements u' of F' are the same as the pairs (u_1, u_2) of I-summable U-elements of F which satisfy the condition $u_i \gg \theta_F$, i = 1, 2. By a similar but simpler proof, the L-elements of F' are all summable and consist of all pairs (l_1, l_2) of L-elements of F such that $l_i \geqq \theta_F$, i = 1, 2; and $I_1'(l_1, l_2) = (I_1(l_1), I_1(l_2))$.

From this it follows at once that if f_1 and f_2 are in F and $f_i \gg \theta_F (i = 1, 2)$, then $f' = (f_1, f_2)$ has an upper and a lower I'-integral if and only if f_1 and f_2 have upper and lower I-integrals, and in that case $\bar{I}'(f') = (\bar{I}(f_1), \bar{I}(f_2))$ and $\underline{I}'(f') = (\underline{I}(f_1), \underline{I}(f_2))$. In particular, f' is summable if and only if f_1 and f_2 are summable, and then $I'(f') = (I(f_1), I(f_2))$.

For each element $f' = (f_1, f_2)$ of F_+' for which $f_i \gg \theta_F$ (i = 1, 2), we define $\Phi_0(f') = f_1 f_2$; and for each pair (g_1, g_2) in the domain of the multiplication in G in which $g_i \geqq \theta_G$ (i = 1, 2) we define $\Gamma(g_1, g_2) = g_1 g_2$. Then if $e' = (e_1, e_2)$ is in E', by hypothesis $I_0(\Phi_0(e')) = \Gamma(I_0'(e'))$. Now hypotheses (18.1[$\sigma$]) are satisfied. By conclusion (iv) of (18.1[σ]) we find that if f_1 and f_2 are I-summable members of F_+, and $f_i \gg \theta_F$ (e = 1, 2), then $f_1 f_2$ is I-summable, and $I(f_1 f_2) = I(f_1) I(f_2)$.

There exist elements e_1, e_2 of E such that $\theta_F \ll e_1 \ll e_2$, so by (9.7) there is a summable U-element $u^* \gg \theta_F$. This u^* is in F_+, since F_+ is Dedekind [σ]-closed and contains the elements $e_3, e_4, \ldots$ of (9.7).

If f is summable and in F_+, so are -f and θ_F by (19.3), and $\bigvee\{f, -f, \theta_F\}$ is summable by (13.1). Since F_+ is a lattice, $\bigvee\{f, -f, \theta_F\}$ is a summable member of F_+. So is $\bigvee\{f, -f, \theta_F\} - f$; and they are both $\geqq \theta_F$. Hence the elements $f' = \bigvee\{f, -f, \theta_F\} + u^*$ and $f'' = \bigvee\{f, -f, \theta_F\} - f + u^*$ are summable members of F_+, and $f' - f'' = f$, and $f' \gg \theta_F$ and $f'' \gg \theta_F$.

If f_1 and f_2 are summable members of F_+, we can thus write $f_i = f_i' - f_i''$ (i = 1, 2), where $f_i' \gg \theta_F$ and $f_i'' \gg \theta_F$ (i = 1, 2). By the previous proof, all the equations

$$I(f_1' f_2') = I(f_1') I(f_2') ,$$

$$I(f_1' f_2'') = I(f_1') I(f_2'') ,$$

$$I(f_1''f_2') = I(f_1'')I(f_2') ,$$

$$I(f_1''f_2'') = I(f_1'')I(f_2'')$$

are meaningful and correct. Hence by (19.3)

$$I(f_1'f_2' - f_1'f_2'' - f_1''f_2' + f_1''f_2) = (I(f_1') - I(f_1''))(I(f_2') - I(f_2'')) ,$$

or $I(f_1f_2) = I(f_1)I(f_2)$, completing the proof.

§ 20. SUMMABILITY OF PRODUCTS

When a binary multiplication is defined in F_+, but is not related to a binary multiplication in G as in Theorem (19.5), we can still prove something about the summability of products of summable elements, under suitable hypotheses. However, the theorem we establish is little more than a transcription of a well-known theorem concerning Lebesgue integrals.

(20.1) **THEOREM.** Let (19.1[σ]) be satisfied, and also the following hypotheses.

(a) F_+ is a conditionally [σ-]complete lattice embedded in F.

(b) F_+ is a ring embedded in F; the identity-element in F_+ will be called θ_F.

(c) Whenever e_1 and e_2 are in E, so is e_1e_2.

(d) Whenever e_1, e_1', e_2 and e_2' are members of E such that $e_1' \gg e_1 \gg \theta_F$ and $e_2' \gg e_2 \gg \theta_F$, the inequality $e_1'e_2' \gg e_1e_2$ is satisfied.

(e) To each element e of E corresponds a positive number B_e such that if f is in F_+ and $f \geqq \theta_F$, then $ef \leqq B_ef$ and $fe \leqq B_ef$.

Then if f_1 and f_2 are summable elements of F_+, and there exist elements e_1 and e_2 of E such that $|f_1| \leqq e_1$ and $|f_2| \leqq e_2$, the product f_1f_2 is also summable.

(REMARK: Hypothesis (e) does not demand that F_+ be a linear system; it is a group, so nf is defined for f in F_+ and n a positive integer.)

Let us first suppose $f_i \gg \theta_F$ $(i = 1, 2)$. Choose elements

e_1', e_2' of E such that $e_i' \gg e_i$ $(i = 1, 2)$. By (9.7), there are U-elements u_1^*, u_2^* and L-elements l_1^*, l_2^* such that $\theta_F \ll l_i^* \ll f_i$, $e_i \ll u_i^* \ll e_i'$. Let B be a number such that if f is in F_+ and $f \geqq \theta_F$, $e_i'f \leqq Bf$, and $fe_i' \leqq Bf$, $(i = 1, 2)$. Define U_i to be the set $\{u\colon u$ in $U[\geqq f_i]$ and $u \leqq u_i^*\}$, $i = 1, 2$; and define L_i to be the set $\{l\colon l$ in $L[\leqq f_i]$ and $l \geqq l_i^*\}$ $(i = 1, 2)$. Clearly it makes no difference in defining the upper and lower integrals of f_i if we use only U-elements in U_i and L-elements in L_i. If u_i is in U_i and l_i in L_i $(i = 1, 2)$, by the Dedekind closure of F_+ they are in F_+, and

$$\begin{aligned} u_1u_2 - l_1l_2 &= u_1(u_2 - l_2) + (u_1 - l_1)l_2 \\ &\leqq e_1'(u_2 - l_2) + (u_1 - l_1)e_2' \\ &\leqq B(u_2 - l_2 + u_1 - l_1) \ . \end{aligned}$$

Also $u_1u_2 \geqq f_1f_2 \geqq l_1l_2$, hence

$$\begin{aligned} \theta_F &\leqq \overline{I}(f_1f_2) - \underline{I}(f_1f_2) \\ &\leqq I_1(u_1u_2) - I_1(l_1l_2) \\ &\leqq B(I_1(u_2) - I_1(l_2) + I_1(u_1) - I_1(l_1)) \ . \end{aligned}$$

But $\bigwedge I_1(U_i) = \bigvee I_1(L_i) = I(f_i)$ $(i = 1, 2)$, so by (6.8) the infimum of the right member is θ_F. Hence f_1f_2 is summable. It is in F_+, because the binary multiplication maps into F_+.

Now we return to the hypothesis $|f_i| \leqq e_i$, $(i = 1, 2)$. The elements $I(f_i)$ and $I(\theta_F)$ have the common upper bound $I_0(e_i')$, where $e_i' \gg e_i$; so f_i^+ and f_i^- are summable members of F_+ $(i = 1, 2)$. Choose elements e_0, e_0' of E such that $e_0' \gg e_0 \gg \theta_F$, and choose (by (9.7)) a U-element u^* such that $e_0 \ll u^* \ll e_0'$. This is in F_+. Define $h_i = f_i^+ + u^*$, $k_i = f_i^- + u^*$ $(i = 1, 2)$; then $f_i = h_i - k_i$, and also $\theta \ll h_i \ll e_i' + e_0'$ and $\theta \ll k_i \ll e_i' + e_0'$. By the preceding proof, the products $h_1h_2, h_1k_2, k_1h_2, k_1k_2$ are all summable members of F_+. So by (19.3) $h_1h_2 - h_1k_2 - k_1h_2 + k_1k_2$ is summable. But this is f_1f_2.

§ 21. OPERATIONS ON MEASURABLE ELEMENTS

Some of the theorems of Section 19 have almost obvious analogues for measurable elements. The simplest are the following two.

(21.1) THEOREM. Let (19.1[σ]) be satisfied. Let f be a measurable element of F which is in F_+.

Then $-f$ is measurable.

If g and h are summable and $g \leqq h$, we choose a summable L-element $l \leqq g$ and a summable U-element $u \geqq h$. Then $\text{mid}\,(g,h,\text{mid}\,(l,u,-f)) = \text{mid}\,(g,h,-f)$, which must be shown summable, so it is enough to prove $\text{mid}\,(l,u,-f)$ summable. Let S_1 and S_2 be subsets of E associated with l and u respectively. Then

$$\begin{aligned} \text{mid}\,(l,u,-f) &= \bigvee_{e_2 \text{ in } S_2} \Big[\bigwedge_{e_1 \text{ in } S_1} \text{mid}\,(e_1,e_2,-f)\Big] \\ &= \bigvee_{e_2 \text{ in } S_2} \Big[\bigwedge_{e_1 \text{ in } S_1} (-\text{mid}\,(-e_1,-e_2,f))\Big] \\ &= -\bigwedge_{-e_2' \text{ in } S_2} \Big[\bigvee_{-e_1' \text{ in } S_1} \text{mid}\,(e_1',e_2',f)\Big] \\ &= -\,\text{mid}\,(-l,-u,f)\ , \end{aligned}$$

which is summable by (19.3) and the definition of measurability.

(21.2) **THEOREM.** Let the hypotheses of Theorem (19.4) be satisfied. Let f be a measurable element which belongs to F_+, and let c be a real number. Then cf is measurable.

As in the preceding proof, we need only show that $\text{mid}\,(l,u,cf)$ is summable when l is a summable L-element and u a summable U-element and $u \geqq l$. Also, by virtue of (21.1) it is enough to consider $c > 0$. Let S_1, S_2 be associated with l,u respectively, and let $c^{-1}S_i$ be the set $\{c^{-1}e\colon e \text{ in } S_i\}$ $(e = 1, 2)$. Then $\bigwedge c^{-1}S_1$ is a summable L-element which we call l_c, and $\bigvee c^{-1}S_2$ is a summable U-element which we call u_c, and

$$\begin{aligned} \text{mid}\,(l,u,cf) &= \bigvee_{e_2 \text{ in } S_2} \Big[\bigwedge_{e_1 \text{ in } S_1} \text{mid}\,(e_1,e_2,cf)\Big] \\ &= \bigvee_{e_2' \text{ in } c^{-1}S_2} \Big[\bigwedge_{e_1' \text{ in } c^{-1}S_1} \text{mid}\,(ce_1',ce_2',cf)\Big] \\ &= c \bigvee_{e_2' \text{ in } c^{-1}S_2} \Big[\bigwedge_{e_1' \text{ in } c^{-1}S_1} \text{mid}\,(e_1',e_2',f)\Big] \\ &= c\ \text{mid}\,(l_c,u_c,f)\ . \end{aligned}$$

This last is summable by (19.4).

To establish the measurability of the sum of measurable elements it seems at least desirable and probably necessary to add a hypothesis, assuming a strengthened form of normality for F. Let us first introduce a definition.

(21.3) **DEFINITION.** If F_+ is a group embedded in a partially ordered set F, $R_{F,+}$ is defined to be the set of all smoothly rising extended real-valued functions on F which are finite on F_+ and satisfy the equation $r(f_1 + f_2) = r(f_1) + r(f_2)$ whenever f_1 and f_2 are in F_+.

(21.4) **THEOREM.** Let the hypotheses of (19.3[σ]) hold. Assume further

(i) if f_1 and f_2 are distinct elements of F, there exists a function r in $R_{F,+}$ such that $r(f_1) \neq r(f_2)$, and

(ii) if f is in F and $r(f)$ is finite for all r in $R_{F,+}$, then f is in F_+.

Then whenever f and g are measurable members of F_+, so is $f + g$.

As before, it is enough to show that when l is a summable L-element and u a summable U-element and $l \leqq u$, the element mid $(l,u,f + g)$ is summable. Let S' and S'' be sets associated with l and u respectively. Let S_m' be the set of all elements of the form $e_0' - (e_1'' - e_1') - \dots - (e_m'' - e_m')$, where $e_0', e_1', \dots, e_m'$ are in S' and $e_1'', \dots, e_m''$ are in S''; let S_n'' be the set of all elements of the form $e_0'' + (e_1'' - e_1') + (e_n'' - e_n')$, where $e_0'', e_1'', \dots, e_n''$ are in S'' and $e_1', \dots, e_n'$ are in S'.

We easily verify that S_m' is directed by $\ll$ and S_n'' by $\gg$, and that $l_m \equiv \bigwedge S_m'$ is a summable L-element and $u_n \equiv \bigvee S_n''$ is a summable U-element. Also, if $e_0', \dots, e_m'$ are in S' and $e_1'', \dots, e_m''$ in S'', for each r in $R_{F,+}$ we have

$$r(e_0' - [e_1'' - e_1'] - \dots - [e_m'' - e_m']) = r(e_0) - r(e_1'') + r(e_1') - \dots - r(e_m'') + r(e_m') ,$$

and since r is smoothly rising this implies that

$$r(l_m) = r(l) - m[r(u) - r(l)] ,$$

where $r(u) - r(l)$ is $\geqq 0$ and may be $+\infty$, in which case $r(l_m) = -\infty$. Likewise $r(u_n) = r(u) + n[r(u) - r(l)]$

We now define $f_{m,n} = \text{mid}\ (l_m,u_n,f)$, $g_{m,n} = \text{mid}\ (l_m,u_n,g)$. Since r is isotone,

$$r(f_{m,n}) = \text{mid}\ (r(l_m),r(u_m),r(f)) \ ,$$

and analogously for $r(g_{m,n})$. But $r(u_n)$ can never be $-\infty$, nor can $r(l_n)$ be $+\infty$, and $r(f)$ is finite, so $r(f_{m,n})$ is finite. By hypothesis, $f_{m,n}$ is in F_+; and similarly, so is $g_{m,n}$.

Since F is a complete lattice, the limit

$$s = \lim_{n\to\infty} \ (\lim_{m\to\infty} [f_{m,n} + g_{m,n}])$$

exists. For r in $R_{F,+}$, we have

$$(*) \qquad r(s) = \lim_{n\to\infty} \ (\lim_{m\to\infty} [r(f_{m,n}) + r(g_{m,n})]) \ .$$

We now distinguish two cases. If $r(u) - r(l) > 0$, for all large m and n we have

$$r(u_n) > r(f) > r(l_n) \quad \text{and} \quad r(u_n) > r(g) > r(l_n) \ ,$$

so $r(f_{m,n}) = r(f)$ and $r(g_{m,n}) = r(g)$. Then by (*) we find

$$r(s) = r(f) + r(g) = r(f + g) \ ,$$

whence

$$(**) \qquad r(\text{mid}\ (l,u,s)) = r(\text{mid}\ (l,u,f + g)) \ .$$

If on the other hand $r(u) - r(l) = 0$, then $\text{mid}\ (r(u),r(l),z) = r(u)$ for all real numbers z, and (**) continues to hold. Thus (**) holds for all r in $R_{F,+}$, whence

$$\text{mid}\ (l,u,s) = \text{mid}\ (l,u,f + g) \ .$$

But

$$\lim_{m\to\infty} \ \lim_{n\to\infty} \text{mid}\ (l,u,f_{m,n} + g_{m,n})$$
$$= \text{mid}\ (l,u,s) \ ,$$

and the elements $\text{mid}\ (l,u,f_{m,n} + g_{m,n})$ are summable and are between l and u. Hence by (15.1) $\text{mid}\ (l,u,s)$ is summable; so $\text{mid}\ (l,u,f + g)$ is summable, and $f + g$ is measurable.

CHAPTER V

REAL-VALUED FUNCTIONS

§ 22. INTEGRALS OF REAL-VALUED FUNCTIONS

An important special case of the preceding theory is that in which F consists of all extended-real-valued functions on a domain T and F_+ consists of all (finite) real-valued functions on T, the definition of order, addition and multiplication being the standard ones. These sets F and F_+ satisfy all the requirements placed on sets F and F_+ in preceding theorems, so (with appropriate hypotheses on G, E, I_0, etc.) all preceding theorems are available. However, one new idea enters, namely the concept of a measurable set, and this we must investigate. We do not wish to restrict G to be the real number system, but we do wish to perform linear operations in G. Accordingly, we assume the following postulates.

(22.1[σ]) POSTULATES.

(a) F consists of all extended-real-valued functions on a domain T; and for all f_1, f_2 on F, $f_1 \geqq f_2$ means $f_1(t) \geqq f_2(t)$ for all t in T.

(b) F_+ consists of all (finite-) real-valued functions on T; if f_1 and f_2 are in F_+, $f_1 + f_2$ is the function $(f_1(t) + f_2(t): t \text{ in } T)$, and scalar multiplication and binary multiplication are analogously defined.

(c) $\gg$ is a strengthening of $\geqq$.

(d) G is a normal Dedekind-complete partially ordered linear system.

(e) I_0 is an isotone function whose domain is a subset E of F and whose range is contained in G.

(f) For each [countable] subset S of E

directed by $\ll$ and such that $\bigwedge S \leqq 0$ and $I_0(S)$ is bounded below, $\bigwedge I_0(S) \leqq \theta_G$.

(g) If e_1 and e_2 are in E, and $I_0(e_1)$ and $I_0(e_2)$ have a common upper or lower bound in G, there exist elements e' and e'' of E such that $e' \ll e_i \ll e''$, $i = 1, 2$.

(h) If e_1, e_2 and e_3 are in E, and $I_0(e_1)$ and $I_0(e_2)$ have a common upper or lower bound in G, then for every f in F such that $f \gg \text{mid}\,(e_1,e_2,e_3)$ there is an e in E such that $f \gg e \gg \text{mid}\,(e_1,e_2,e_3)$; and dually.

(i) Whenever e_1 and e_2 are in E and c_1 and c_2 are real numbers, $c_1e_1 + c_2e_2$ is in E, and $I_0(c_1e_1 + c_2e_2) = c_1I_0(e_1) + c_2I_0(e_2)$; and if e_1, e_1', e_2 and e_2' are in E and $e_i \ll e_i'$ $(i = 1, 2)$ and $c < 0$, then $e_1 + e_2 \ll e_1' + e_2'$ and $ce_1 \gg ce_1'$.

It is obvious that if we assume (22.1[σ]) we at once obtain the conclusions of (19.3) and (19.4); and if we also assume hypotheses (19.5d,e,f) we obtain the conclusions of (19.5). However, we can establish stronger statements. As usual, let us define $\infty + x = x + \infty = \infty$ if $x > -\infty$, $-\infty + x = x + (-\infty) = -\infty$ if $x < +\infty$. Then we can define $f_1 + f_2$ to be $(f_1(t) + f_2(t)\colon t \text{ in } T)$ provided only that there is no t in T at which one of the values $f_1(t)$, $f_2(t)$ is ∞ and the other is $-\infty$. This agrees with the definition of + when f_1 and f_2 are in F_+. In particular, no U-function is ever $-\infty$, and no L-function is ever $+\infty$, so $f_1 + f_2$ is defined whenever f_1 and f_2 are both U-functions or both L-functions.

(22.2) **THEOREM.** Let (22.1[σ]) hold. Let f_1 and f_2 be functions such that $\bar{I}(f_i)$ and $\underline{I}(f_i)$ are defined $(i = 1, 2)$, and let s be a function such that $s(t) = f_1(t) + f_2(t)$ whenever the sum is defined, s(t) being arbitrary (finite, $+\infty$ or $-\infty$) elsewhere in T. Then

$$\underline{I}(f_1) + \underline{I}(f_2) \leqq \underline{I}(s) \leqq \bar{I}(s) < \bar{I}(f_1) + \bar{I}(f_2) ,$$

$$\bar{I}(-f_1) = -\underline{I}(f_1),\ \underline{I}(-f_1) = -\bar{I}(f_1) .$$

In particular, if f_1 and f_2 are summable, so are

s and $-f_1$, and $I(s) = I(f_1) + I(f_2)$ and $I(-f) = -I(f_1)$.

If u_i is in $U[\geqq f_i]$ $(i = 1, 1)$, then $u_1(t) + u_2(t)$ is defined for all t in T. If S_1, S_2 are associated with u_1, u_2 respectively, then $u_1 + u_2 = \bigvee\{e_1 + e_2 \colon e_1$ in S_1 and e_2 in $S_2\}$, so $u_1 + u_2$ is a U-function and

$$\begin{aligned} I_1(u_1 + u_2) &= \bigvee\{I_0(e_1 + e_2) \colon e_1 \text{ in } S_1 \text{ and } e_2 \text{ in } S_2\} \\ &= \bigvee I_0(S_1) + \bigvee I_0(S_2) \\ &= I_1(u_1) + I_1(u_2) \ . \end{aligned}$$

Whenever $f_1(t) + f_2(t)$ is defined, we have

$$s(t) = f_1(t) + f_2(t) \leqq u_1(t) + u_2(t) \ .$$

Where $f_1(t) + f_2(t)$ is undefined, one summand (say $f_1(t)$) is $+\infty$. Then $u_1(t)$ is $+\infty$, while $u_2(t)$ cannot be $-\infty$, so $u_1(t) + u_2(t) = +\infty \geqq s(t)$. Hence $u_1 + u_2 \geqq s$, and

$$\bar{I}(s) \leqq \bar{I}_1(u_1 + u_2) = I_1(u_1) + I_1(u_2) \ .$$

This holds for all u_1, u_2 in $U[\geqq f_1]$, $U[\geqq f_2]$ respectively, so

$$\begin{aligned} \bar{I}(s) &\leqq \bigwedge I_1(U[\geqq f_1]) + \bigwedge I_1(U[\geqq f_2]) \\ &= \underline{I}(f_1) + \underline{I}(f_2) \ . \end{aligned}$$

The statement concerning $\underline{I}(s)$ is established dually, and those concerning $-f_1$ are easy to establish.

(22.3) **THEOREM**. Let (22.1[σ]) hold. If f is summable, and $I(f)$ and θ_G have a common upper or lower bound, $|f|$ is also summable and

$$-I(|f|) \leqq I(f) \leqq I(|f|) \ .$$

For then by (22.2) and (6.7), $I(-f)$, θ_G and $I(f)$ have a common bound, and by (13.1) $|f|$ is summable. By (11.8) the inequalities hold.

For each fixed t in T, the function $r_t = (r_t(f) = f(t) \colon t$ in $T)$ is smoothly rising on F and is finite and additive on F_+, so it belongs to $R_{F,+}$. It is obvious that hypotheses (i) and (ii) of (21.3) hold, so by this and (21.1) and (21.2) we have

(22.4) COROLLARY. Let (22.1[σ]) hold. If f_1 and f_2 are finite-valued and measurable and c is a real number, then $f_1 + f_2$ and cf_1 are also measurable.

REMARK. This can be improved; if f_1 and f_2 are measurable and $f_1(t) + f_2(t)$ is defined for all t in T then $f_1 + f_2$ is measurable. The proof is essentially that of (21.3); the chief change is that $f_{m,n}$ and $g_{m,n}$ may not be finite-valued, but in any case their sum is defined.

Given any set M, we shall denote by χ_M its characteristic function, that is the function whose value is 1 at each point of M and is 0 elsewhere.

(22.5) DEFINITION. If M is a subset of D, M has finite measure if χ_M is summable, and in this case the measure of M is

$$mM = I(\chi_M) ;$$

the set M is a measurable set if χ_M is a measurable element of F.

If for a sequence $E_1, E_2, \ldots$ of sets we denote by $\limsup E_m$ the set of all points which belong to infinitely many E_n and by $\liminf E_n$ the set of points which belong to all but finitely many of the E_n, we find that the characteristic function of $\limsup E_n$ is the upper limit of the characteristic functions of the E_n, and likewise for the lower limit.

(22.6) COROLLARY. Let (22.1[σ]) hold, and let $E_1, E_2, \ldots$ be measurable sets. Then $\bigcup_n E_n$, $\bigcap_n E_n$, $\limsup E_n$ and $\liminf E_n$ are measurable sets.

For $\chi_{\bigcup_n E_n} = \bigvee_n \chi_{E_n}$, and dually; whence the conclusion follows by (17.8).

(22.7) COROLLARY. Let (22.1[σ]) hold. The empty set has measure θ_G; and for every set E of finite measure, $mE \geqq \theta_G$.

Let f be any summable member of F_+. The characteristic function of the empty set is $f - f$, and $I(f - f) = \theta_G$ by (19.3). The other statement follows from this and (11.8).

(22.8) COROLLARY. Let (22.1[σ] hold. If E_1 has finite measure, and E_2 is measurable and contained in E_1, then E_2 has finite measure, and $\theta_G \leqq mE_2 \leqq mE_1$.

This follows from (17.2), (22.5) and (11.8).

(22.9) COROLLARY. Let (22.1[σ]) hold. If $E_1 \subset E_2$ and $mE_2 = \theta_G$, then E_1 has measure θ_G.

Since $0 \leqq \chi_{E_1} \leqq \chi_{E_2}$, this follows from (11.4) and (11.5).

(22.10) COROLLARY. Let (22.1[σ]) hold. If E_1 and E_2 have finite measure, so have $E_1 \cup E_2$ and $E_1 \cap E_2$, and $m(E_1 \cup E_2) + m(E_1 \cap E_2) = mE_1 + mE_2$. In particular, if E_1 and E_2 are disjoint, $m(E_1 \cup E_2) = mE_1 + mE_2$.

Since $\chi_{E_1 \cup E_2} = \chi_{E_1} \vee \chi_{E_2}$ and dually, the first statement follows from (6.3), (13.1) and (19.3); the second statement follows from this and (22.7).

(22.11) COROLLARY. Let (22.1[σ]) hold. Let $E_1, E_2, \ldots$ be sets of finite measure. If all E_n are contained in a set of finite measure, or else if (14.1) holds and $\Sigma\, m\, E_n$ converges, the set $\bigcup_n E_n$ has finite measure, and its measure is $\leqq \Sigma\, m\, E_n$ if this exists; equality holds if the E_n are disjoint.

For each j the set $S_j = \bigcup_{n=1}^{j} E_n$ has finite measure $\leqq \Sigma_1^j\, mE_n$, by (22.10), equality holding if the E_n are disjoint. If all E_n are contained in a set of finite measure, by (14.4) $\bigcup_j S_j$ has finite measure, and $m(\bigcup_j S_j) = \lim_{j \to \infty} mS_j \leqq \Sigma mE_n$ if this exists; equality holds when the E_n are disjoint. If (14.1) holds, the desired conclusion follows from (14.3).

(22.12) COROLLARY. Let (22.1[σ]) hold. Let $E_1, E_2, \ldots$ be sets having $mE_n = \theta_G$, $n = 1, 2, \ldots$. If all E_n

are contained in a set of finite measure, or if (14.1) holds, the union $\bigcup_n E_n$ has measure θ_G.

This follows from (22.11) and (22.7).

By an obvious extension of the usual phrase, "almost everywhere" shall mean "except in a set measure θ_G."

(22.13) **THEOREM**. If (22.1[σ]) holds and f is summable, it is finite almost everywhere in T.

Let N be the set on which $f(t) = \pm\infty$. The hypotheses of (22.2) are satisfied with $f_1 = f$, $f_2 = -f$, $s = \chi_N$, so χ_N is summable, and $mN = I(f) + I(-f) = \theta_G$.

(22.14) **DEFINITION**. Let (22.1[σ]) hold. Two members f_1, f_2 of F are equivalent if $f_1(t) = f_2(t)$ almost everywhere.

(22.15) **THEOREM**. Let (22.1[σ]) hold. Let f_1 be a function such that $\underline{I}(f_1)$ and $\overline{I}(f_1)$ exist, and let f_2 be equivalent to f_1. Then if either

(a) there exist summable members f', f'' of F such that $f' \leqq f_i \leqq f''$ $(i = 1, 2)$, or else

(b) (14.1) holds,

it follows that $\overline{I}(f_2) = \overline{I}(f_1)$ and $\underline{I}(f_2) = \underline{I}(f_1)$; in particular, if f_1 is summable so is f_2, and $I(f_2) = I(f_1)$.

First let us suppose that (a) holds. Let N be the set $\{t\colon t \text{ in } T \text{ and } f_1(t) \neq f_2(t)\}$; then by hypothesis $mN = \theta_G$. Since f' and f'' are summable, we can find a summable L-function $l \leqq f'$ and a summable U-function $u \geqq f''$; and $u - l$ is defined and is a non-negative summable U-function. Let h be the function which has the value $u(t) - l(t)$ for each t in N and is 0 elsewhere. For each positive integer n, $I(n\chi_N) = nI(\chi_N) = n\theta_G = \theta_G$, so using (13.1), $\theta_G \leqq I([n\chi_N] \wedge [u - l]) \leqq I(n\chi_N) = \theta_G$. By (14.4), $I(h) = \lim I([n\chi_n] \wedge [u - l]) = \theta_G$. Let s be the function which has the value $f_1(x) + h(x)$ whenever this is defined and the value $u(x)$ elsewhere. By (22.2),

$$\overline{I}(s) \leqq \overline{I}(f_1) + \overline{I}(h) = \overline{I}(f_1) .$$

But where $f_1(t) + h(t)$ is defined we have either t not in N, in which case $f_2(t) = f_1(t) = f_1(t) + h(t)$, or else t in N, in which case

$f_2(t) \leqq f_1(t) + [u(t) - l(t)] = f_1(t) + h(t)$; and where $f_1(t) + h(t)$ is undefined we have $f_2(t) \leqq u(t) = s(t)$. So $f_2 \leqq s$, and $\overline{I}(f_2) \leqq \overline{I}(f_1)$. Interchanging the roles of f_1 and f_2 establishes the reverse inequality, so $\overline{I}(f_2) = \overline{I}(f_1)$. The other equation is established dually.

If hypothesis (b) holds, let h be the function which is ∞ on N and 0 elsewhere. By (14.3), $I(h) = \lim I(n\chi_N) = \theta_G$, so there is a summable U-function $u_1 \geqq h$. If u is in $U[\geqq f_1]$ and l is in $L[\leqq f_1]$, then $l - u_1 \leqq f_i \leqq u + u_1$ $(i = 1, 2)$, and hypothesis (a) is verified.

(22.16) THEOREM. Let (22.1[σ]) hold. Let f and g be summable members of F such that $g \geqq f$ and $I(g) = I(f)$. Then f and g are equivalent.

Let N be the subset of T on which f and g are different. Let d be the function which is equal to $g(t) - f(t)$ for t on N and to 0 elsewhere. Then $d \geqq 0$, and $I(d) = \theta_G$. For each positive integer k let N_k be the set $(t: t$ in T and $d(t) \geqq 1/k)$. Then $0 \leqq (1/k)\chi_{N_k} \leqq d$, so $I([1/k]\chi_{N_k}) = \theta_G$, and $mN_k = \theta_G$. Since $N = \bigcup_k N_k$, $mN = \theta_G$, and f and g are equivalent.

§ 23. MEASURABLE FUNCTIONS AND LEBESGUE LADDERS

A function ϕ on a vector space is called positively homogeneous (of degree 1) if $\phi(kx) = k\phi(x)$ when $k \geqq 0$.

(23.1) THEOREM. Let (22.1[σ]) hold. Let $\phi = (\phi(y_1, \ldots, y_n): y_i$ real, $i = 1, \ldots, n)$ be a function of Baire defined and positively homogeneous on n-space. Let $f_1, \ldots, f_n$ be finite-valued measurable functions. Then the function $\phi(f) \equiv (\phi(f_1(t), \ldots, f_n(t)): t$ in $T)$ is measurable

By (17.7), (19.3) and (19.4), if ϕ is a homogeneous linear combination of the y_i or is a lattice combination of such linear combinations, the conclusion holds. By the Stone-Weierstrass theorem [Stone 2], on the set $\Sigma |y_i| = 1$ every continuous function can be uniformly approximated by such lattice combinations. Hence if ϕ is positively homogeneous and continuous, we can find a sequence of lattice-combinations $\phi_1, \phi_2 \ldots$ such that $\lim_n \phi_n(f(t)) = \phi(f(t))$ for all t; and by (17.8) $\phi(f)$ is measurable. The extension to Baire functions is immediate.

(23.2) THEOREM. Let (21.1[σ]) hold and let the whole space T be a measurable set. Let $\phi = (\phi(y_1, \ldots, y_n): y_i$ real, $i = 1, \ldots, n)$ be a Baire function on n-space. If $f_1, \ldots, f_n$ are finite-valued and measurable, $\phi(f_1, \ldots, f_n)$ is measurable.

If ϕ is a linear function $a + b_1y_1 + \cdots + b_ny_n$, the conclusion is valid by (22.4), since the function identically 1 is measurable. All lattice combinations of such functions are measurable, by (17.8). By the Stone-Weierstrass theorem, if ϕ is continuous for each positive integer m it can be approximated to within $1/m$ on the interval $\{y : -m \leqq y_i \leqq m, i = 1, \ldots, n\}$ by such a combination ϕ_m. Hence $\lim_{n\to\infty} \phi_m(f(t)) = \phi(f(t))$ for all t, and $\phi(f)$ is measurable. The set of all functions ϕ for which $\phi(f)$ is measurable contains all continuous functions and is closed under passage to the limit, so it contains all Baire functions.

(23.3) THEOREM. Let (22.1[σ]) hold, and let f be a finite-valued function in F. If, for each interval (a,b] which does not contain 0, the set $\{t:$ t in T and $a < f(t) \leqq b\}$ is measurable, then f is measurable. Conversely, if T is a measurable set and f is a measurable function, then for each interval (a,b] the set $\{t:$ t in T and $a < f(t) \leqq b\}$ is measurable.

Let m be a positive integer, and let $J_{m,n}$ be the interval $((n - 1)/m, n/m]$, $n = 0, \pm 1, \pm 2, \ldots$. For each n except perhaps 0, the set $E_{m,n} = \{t: f(t)$ in $J_{m,n}\}$ is measurable. The series $\Sigma_{-\infty}^{\infty} n\chi_{E_{m,n}}$ converges at each point, having in fact at most one non-zero term, so by (22.4) and (17.8) its sum is measurable. This sum nowhere differs more than $1/m$ from $f(t)$, so if we let m increase and apply (17.8) we see that f is measurable.

Conversely, let ϕ be the characteristic function of (a,b]. This is a Baire function, so by (23.2) $\phi(f)$ is measurable. But $\phi(f)$ is the characteristic function of the set $\{t:$ t in T and $a < f(t) \leqq b\}$.

(23.4) DEFINITION. By a "Lebesgue ladder" we shall mean a pair of sequences of real numbers $(a_n : n = 0, \pm 1, \pm 2, \ldots)$ and $(b_n : n = 0, \pm 1, \pm 2, \ldots)$ with the properties

(a) The numbers $a_n - a_{n-1}$ are positive and bounded.

(b) For each n, $a_{n-1} \leqq b_n \leqq a_n$.

(c) $\lim_{n \to \infty} a_n = \infty$, $\lim_{n \to \infty} a_{-n} = -\infty$.

By the "mesh" of the ladder we shall mean the supremum of the numbers $a_n - a_{n-1}$, $n = 0, \pm 1, \pm 2, \ldots$.

If we try to follow the familiar "ladder" method of defining the integral with the help of the measure, we find a difficulty when T lacks finite measure that can be overcome, but by a device which we would prefer to avoid using when T has finite measure. We succeed in treating each case in the desired manner by the following artifice. Let Λ be a Lebesgue ladder. If T has finite measure, $N(\Lambda)$ shall mean the set of all integers. If T does not have finite measure, $N(\Lambda)$ shall consist of all those integers n such that the distance from 0 to the interval $[a_{n-1}, a_n]$ is at least equal to the length $a_n - a_{n-1}$ of the interval. A summation designated by a sign Σ' shall mean a summation over some subset of $N(\Lambda)$.

Let f be in F. If for each n in $N(\Lambda)$ the set $\{t : a_{n-1} < f(t) \leqq a_n\}$ has finite measure, and the finite sums

$$\sum_{n=-h}^{\prime\, k} b_n \, m\{t : a_{n-1} < f(t) \leqq a_n\}$$

converge to a limit as $h, k \to \infty$, this limit is called the "Lebesgue sum" associated with f and Λ, and is designated by $\sigma(f;\Lambda)$:

$$\sigma(f;\Lambda) = \sum_{n=\infty}^{\prime\, \infty} b_n \, m\{t : a_{n-1} < f(t) \leqq a_n\} .$$

With the help of these concepts we can relate our definition to the integral in the manner of Lebesgue.

(23.5) **THEOREM**. Assume that (22.1[σ]) holds, that F is the set of all extended real-valued functions on T, that the set T is measurable, and that f and $|f|$ are summable. Then:

(i) if Λ is a Lebesgue ladder, the Lebesgue sum $\sigma(f;\Lambda)$ is defined;

(ii) if $\Lambda_1, \Lambda_2, \ldots$ is a sequence of Lebesgue ladders with norms approaching 0, $\sigma(f;\Lambda_n)$ converges to $I(f)$.

Let Λ be a Lebesgue ladder, with notation (23.4). Let E_n be the set $\{t : a_{n-1} < f(t) \leqq a_n\}$; by (23.3), this is measurable. So if h and k are positive integers, the function

$$s_{h,k} = \sum\nolimits'^{k}_{n=-h} b_n \chi_{E_n}$$

is measurable. We now separate two cases. If mT is finite, we define $g = |f| + \text{mesh}\ \Lambda$. This is summable. We readily verify that for each t in T the inequality $|s_{h,k}(t)| \leqq g(t)$ holds. If T does not have finite measure, we define $g = 2|f|$; this is obviously summable. Each t such that $s_{h,k}(t) \neq 0$ is in some E_n with n in $N(\Lambda)$; then both b_n and $f(t)$ are in the same interval $[a_{n-1}, a_n]$, and by definition of $N(\Lambda)$ the ratio of b_n to $f(t)$ is between $1/2$ and 2 inclusive. So $|s_{h,k}(x)| \leqq 2|f(t)|$, and therefore $|s_{h,k}| \leqq g$. Now in either case, as h and k tend to ∞, $s_{h,k}$ tends everywhere to

$$\sum\nolimits'^{\infty}_{n=-\infty} b_n \chi_{E_n} .$$

Hence this sum is summable, and its integral is the limit of the integrals $I(s_{h,k})$. But $I(s_{h,k}) = \Sigma' b_n mE_n$, the sum being taken over the set of n in $N(\Lambda)$ for which $-h \leqq n \leqq k$. Hence the Lebesgue sum converges, and its value is the integral of $\Sigma' b_n \chi_{E_n}$. This establishes (i), and incidentally also establishes the unconditional convergence of the series defining the Lebesgue sum.

Next let $\Lambda_1, \Lambda_2, \ldots$ be a sequence of Lebesgue ladders with norms tending to 0. Let b be an upper bound for their norms. Let the ladder Λ_k consist of the sequences $a_{n,k}$, $b_{n,k}$, $n = 0, \pm 1, \pm 2, \ldots$, and let $E_{n,k}$ be the set $\{t : a_{n-1,k} < f(t) \leqq a_{n,k}\}$. Let s_k be the function

$$s_k = \sum_{n \text{ in } N(\Lambda_k)} b_{n,k} \chi_{E_{n,k}} .$$

As k increases, $s_k(t)$ converges to $f(t)$ wherever $f(t)$ is finite, while $s_k(t)$ remains 0 wherever $f(t)$ is $\pm\infty$. Since f is finite except on a set of measure 0, $\lim s_k$ is equivalent to f. If T has finite measure, we define $g = |f| + b$; if T does not have finite measure, we define $g = 2|f|$. In either case, as shown in the preceding paragraph, $|s_k| \leqq g$, and g is summable. So by (22.15) we have $I(\lim s_k) = I(f)$. By (15.2), $I(\lim s_k) = \lim I(s_k) = \lim \sigma(f;\Lambda_k)$, and the proof is complete.

In preparation for our next theorem we establish a lemma.

(23.6) LEMMA. Let (22.1[σ]) hold. Assume also that

(i) for every summable function g, $g \vee 0$ is

the limit of a sequence of summable functions $0 \leqq g_1 \leqq g_2 \leqq \cdots$.

Let f be a function non-negative on T. In order that f be measurable, it is necessary and sufficient that $f \wedge g$ be summable whenever g is summable and non-negative.

Before beginning the proof, we observe that hypothesis (i) holds if either E or G is directed by $\geqq$ or by $\leqq$ (cf. (13.3)) or if T has finite measure, or is the union of countably sets of finite measure.

Suppose f measurable and $g \geqq 0$ and summable. Since 0 is summable, so is mid $(f,0,g)$, which is $f \wedge g$. Conversely, assume that $f \geqq 0$ and that $f \wedge g$ is summable for all summable $g \geqq 0$. Let g and $h \leqq g$ be summable. With the g_n of hypothesis (i), $f \wedge g_n$ is summable for each n, so $f \wedge (g \vee 0)$ is the limit of measurable functions, hence measurable. Also, $g \wedge 0$ is measurable by (17.7), so the sum $f \wedge (g \vee 0) + g \wedge 0 = f \wedge g$ is measurable. So too is $(f \wedge g) \vee h$, which is mid (f,h,g). Since $h \leqq$ mid $(f,h,g) \leqq g$, mid (f,h,g) is summable by (17.2), and f is measurable.

By using Lebesgue ladders we can develop a test for the measurability of sets.

(23.7) **THEOREM**. Let (22.1[σ]) and hypothesis (i) of (23.6) hold, and let T be measurable. If E is a subset of T, it is measurable if and only if for every set E_1 of finite measure, $E \cap E_1$ is also of finite measure.

Assume E measurable. If E_1 has finite measure, its characteristic function is non-negative and summable. By (23.6) $\chi_E \wedge \chi_{E_1}$ is summable. This is the characteristic function of $E \cap E_1$. Conversely, assume that $E \cap E_1$ has finite measure whenever E_1 has. For each positive integer k we construct the Lebesgue ladder Λ_k with $a_{n,k} = b_{n+1,k} = n/k$, $n = 0, \pm 1, \pm 2, \ldots$. If g is non-negative and summable, each set $E_{n,k} = \{t : a_{n-1,k} < g(t) \leqq a_{n,k}\}$ is measurable, by (23.3). Those with $n < 0$ are empty, and 0 and 1 are not in $N(\Lambda_k)$. The sum $s_k = \Sigma_{n=2}^{\infty} b_n \chi E_{n,k}$ is a function which is non-negative and $\leqq g$, and is measurable. Therefore it is summable. In particular, each $E_{n,k}$ has finite measure $(n \geqq 2)$ and the union of all sets on which $b_n \geqq 1$ also has finite measure. Since

$$s_k \wedge \chi_E = \sum_{n=2}^{k} b_n \chi_{E \cap E_n} + \chi_{E \cap [\bigcup_{n>k} E_n]}$$

it is summable by hypothesis. As k increases, $s_k \wedge \chi_E$ converges everywhere to $g \wedge \chi_E$, being between the summable functions 0 and g, so the limit is also summable. By (23.6), this implies that χ_E is measurable, and so E is a measurable set.

If we wish to develop a theory of exterior measure, we are handicapped by the fact that G may not contain a greatest element. If it lacks one, we adjoin an ideal element ∞ such that $\infty \geqq g$ for all g in G and $\infty + g = \infty$ for all g in G. Then for each subset E of T we define the exterior measure m^*E to be $\bar{I}(\chi_E)$ if summable U-functions $u \geqq \chi_E$ exist; otherwise $m^*E = \infty$. We readily prove, by use of (22.2), that $m^*(E_1 \cup E_2) \leqq m^*E_1 + m^*E_2$ for all sets E_1, E_2 contained in T. Also, if m^*E is not ∞ it is the infimum of mE' for all sets E' of finite measure which contain E.

Starting with the exterior measure, we can construct a theory of measure according to Carathéodory's procedure. The relationship with our theory is as follows.

(23.8) **THEOREM.** Assume that (22.1[σ]) and hypothesis (i) of (23.6) hold, that T is measurable, and that E is a subset of T. If E is measurable, for every subset E' of T the equation

$$(*) \qquad m^*(E' \cap E) + m^*(E' - E) = m^*E'$$

is satisfied; and if (*) holds whenever E' has finite measure, E is measurable.

As already remarked, the left member of (*) is $\geqq$ the right. To prove the first conclusion, it remains to show that if E is measurable the reverse inequality holds. We need consider only the case $m^*(E') \neq \infty$. Since E and 1 are measurable, χ_E and $1 - \chi_E$ are measurable. Let u be a summable U-function $\geqq \chi_{E'}$. By (17.8) and (17.2) the integrals $I(u \wedge \chi_E)$ and $I(u \wedge [1 - \chi_E])$ exist; their sum is $I(u)$. But

$$\chi_{E' \cap E} = \chi_{E'} \wedge \chi_E \leqq u \wedge \chi_E ,$$

and similarly

$$\chi_{E'-E} = \chi_{E'} \wedge [1 - \chi_E] \leqq u \wedge [1 - \chi_E] ,$$

so $m^*(E' \cap E) + m^*(E' - E) \leqq I(u)$. From the choice of u this implies that the left member of (*) is $\leqq$ the right member, completing the proof of the first conclusion.

Now assume that (*) holds whenever E' has finite measure. Let E' have finite measure; and to avoid notational complexity, let f', f_1 and f_2 be the characteristic functions of the sets E', $E' \cap E$ and $E' - E$ respectively. Then $f_1 = f' + (-f_2)$, and by (22.2) and (*)

$$\begin{aligned} \underline{I}(f_1) &\geqq \underline{I}(f') - \overline{I}(f_2) \\ &= mE' - m^*(E' - E_2) \\ &= m^*(E' \cap E) \\ &= \overline{I}(f_1) \ . \end{aligned}$$

Hence f_1 is summable, and $E' \cap E$ has finite measure. By (23.7), E is a measurable set.

§ 24. FUBINI'S THEOREM

The form of Fubini's theorem which we shall establish deals with three systems, all satisfying postulates (22.1[•]). In all three systems the set G will consist of the real numbers. The first of the three systems will be distinguished by a prime; thus F' will consist of all extended-real-valued functions on T', and $\gg'$ will be a strengthening of $\geqq$, etc. The second system will be distinguished by a second; thus F'' will consist of all extended-real-valued functions on a set T''. The third system will be distinguished by the absence of a superscript; here F is the set of all extended-real-valued functions on a set T, which we take to be $T' \times T''$.

Under these circumstances Theorem (18.1) provides us with a generalization of Fubini's theorem, as follows.

(24.1) THEOREM. With the notation just explained, assume that the following hypotheses are satisfied.

(H_1) For each e in E and each t' in T', the function $e(t',\cdot) = (e(t',t'') : t''$ in $T'')$ belongs to E'', and if e_1 and e_2 are in E and $e_1 \ll e_2$, then $e_1(t',.) \ll'' e_2(t',\cdot)$.

(H_2) For each e in E, the result of applying I_0'' to e is a function $I_0''(e) = (I_0''[e(t',\cdot)] : t'$ in $T')$ which belongs to E'; and if e_1 and e_2 are in E and $e_1 \ll e_2$, then $I_0''(e_1) \ll' I_0''(e_2)$.

(H_3) For each e in E, $I_0'(I_0''(e)) = I_0(e)$.

Then for every I-summable function f in F:

(C_1) for all t' in T' except those which belong to a set $N' \subset T'$ whose I'-measure is 0, $f(t',\cdot)$ is an I''-summable member of F'';

(C_2) if $I''_*(f)$ is an extended-real-valued function on T' which has the same value as $I''[f(t',\cdot)]$ at all t' in $T' - N'$, then $I''_*(f)$ is I'-summable, and $I'(I''_*(f)) = I(f)$.

To apply (18.1), we choose G' and G to be the real numbers, Γ being the identical mapping. The system $[F,\gg,E,G,I_0]$ of Theorem (18.1) will be the system $[F',\gg',E',G',I_0']$ of the present theorem, and vice versa. For Φ_0 we take I_0''. The hypotheses of (18.1) are then clearly satisfied, $E'_{f'}$ being all of E. Conclusion (iii) becomes $I'(\bar{\Phi}(f)) = I'(\underline{\Phi}(f)) = I(f)$. By (22.16) and (22.13), $\bar{\Phi}(f)$ and $\underline{\Phi}(f)$ are functions on T' which are finite and equal except on a set $N' \subset T'$ whose I'-measure is 0. Let us now consider a point t' in $T' - N'$. From (H_1) it follows readily that if u is a U-function in F and $u \geqq f$, then $u(t',\cdot)$ is a U-function in F'', and $u(t',\cdot) \geqq f(t',\cdot)$. Hence $\bar{I}''[f(t',\cdot)] \leqq \Phi_1(u(t',\cdot))$, and since this holds for all I-summable U-functions $\geqq f$, it implies $\bar{I}''[f(t',\cdot)] \leqq \bar{\Phi}(f)$ at t'. The dual statement also holds, so from the equality of $\bar{\Phi}(f)$ and $\underline{\Phi}(f)$ at t' we deduce that both the upper and lower integrals of $f(t',\cdot)$ also have this same value. That is, on $T' - N'$ the function $f(t',\cdot)$ is I''-summable, and its I''-integral is the same as $\bar{\Phi}(f)$ and $\underline{\Phi}(f)$. This establishes (C_1). It also shows that the function $I''_*(f)$ of (C_2) is equivalent to $\bar{\Phi}(f)$. The real numbers satisfy (14.1), so by (22.15) conclusion (C_2) also holds.

CHAPTER VI

APPLICATIONS

§ 25. MEASURE IN LOCALLY COMPACT SPACES

Let F be the family of all subsets of a locally compact topological space T. We shall use the symbols f^c, f^o to denote the closure and the interior of f, respectively. We assume that E is a non-empty subset of F with the following properties:

(25.1) (i) If e_1 and e_2 are in E, so are $e_1 \cup e_2$, $e_1 \cap e_2$ and $e_1 - e_2$;
(ii) if p is in T and u is an open set containing p, then there exists e in E such that p is in e^o and $e^c \subset u$;
(iii) if e is in E, e^c is compact.

On E we assume a function I_o defined and having the following properties.

(25.2) (i) For all e in E, $0 \leqq I_o(e) < \infty$.
(ii) If e_1 and e_2 are disjoint members of E, $I_o(e_1 + e_2) = I_o(e_1) + I_o(e_2)$.

We define $f_1 \geqq f_2$ to mean $f_1 \supset f_2$, while $f_1 \gg f_2$ shall mean $f_1^o \supset f_2^c$. Let G be the real number system. Then (12.1a,b,c,d) are easily verified and G is normal.

In order to investigate (12.1e) (which is the same as (8.1e)) we first prove a lemma.

(25.3) **LEMMA**. If S is a subset of F directed by $\gg$, then $\bigvee S = \bigvee\{f^o\colon f \text{ in } S\}$; if S is a subset of F directed by $\ll$ then $\bigwedge S = \bigwedge\{f^c\colon f \text{ in } S\}$.

Let S be ordered by $\gg$, and let p belong to $\bigvee S$; then p

belongs to some f in S. Since S is directed by $\gg$, there exists g in S such that $g \gg f$. Then p is in f and $f \subset f^c \subset g^o \subset \bigvee\{f^o\colon f \text{ in } S\}$, so that $\bigvee S \subset \bigvee\{f^o\colon f \text{ in } S\}$. The reverse inclusion is evident, so the first statement is established.

If S is ordered by $\ll$, the set $S' = \{T - f\colon f \text{ in } S\}$ is ordered by $\gg$. Hence, $\bigvee S' = \bigvee\{(T - f)^o\colon f \text{ in } S\} = \bigvee\{T - f^c\colon f \text{ in } S\} = T - \bigwedge\{f^c\colon f \text{ in } S\}$. Since $\bigvee S' = \bigvee\{T - f\colon f \text{ in } S\} = T - \bigwedge\{f\colon f \text{ in } S\} = T - \bigwedge S$, this proves the second statement.

Now let S_1, S_2 be subsets of E directed by $\gg, \ll$ respectively, such that $\bigwedge S_2 \leqq \bigvee S_1$. Each set $e_2^c - e_1^o$ with e_1 in S_1 is compact, by (25.1 iii). No point can belong to all sets $e_2^c - e_1^o$ (e_1 in S_1); hence finitely many such sets $e_{2,j}^c - e_{1,j}^o$ $(j = 1, \ldots, k)$ have an empty intersection. Choose e_2 in S_2 such that $e_2 \ll e_{2j}$ $(j = 1, \ldots, k)$; choose e_1 in S_1 such that $e_1 \gg e_{1,j}$ $(j = 1, \ldots, k)$. Then $e^c_2 - e^o_1$ is contained in all the sets $e^c_{2,j} - e^o_{1,j}$, hence is empty. Consequently $e^c_2 \subset e^o_1$, $e_2 \subset e_1$, and $I_o(e_2) \leqq I_o(e_1)$. Hence $\bigwedge I_o(S_2) \leqq \bigvee I_o(S_1)$, and (8.1e) is established, without the restriction to countable sets S_1.

If k is in F and k is compact, and h is in F, and $k \ll h$, each point of k is by (18.1 ii) interior to an e in E whose closure is in h^o. Finitely many such e have interiors which cover k; their union is an e in E such that $k \ll e \ll h$. Since the empty set and T are in F, this implies (8.1f); and since mid (e_1, e_2, e_3) is in E when e_1, e_2 and e_3 are, this implies (8.1g).

Let e_1, e_2, e_3 belong to E. Write d_1 for $e_1 - e_2 \cup e_3$, d_{12} for $e_1 \cup e_2 - e_3$, and d_{123} for $e_1 \cap e_2 \cap e_3$, the other d_i and d_{ij} being analogously defined. Then $e_1 = d_1 + d_{12} + d_{13} + d_{123}$, and likewise for e_2 and e_3; and also $e_1 \cup e_2 \cup e_3 = d_1 + d_2 + d_3 + d_{12} + d_{13} + d_{23} + d_{123}$, and mid $(e_1, e_2, e_3) = d_{12} + d_{23} + d_{13} + d_{123}$. By (25.1), $\Sigma_{i=1}^{3} I_o(e_i) = I_o(e_1 \cup e_2 \cup e_3) + I_o(\text{mid } (e_1, e_2, e_3)) + I_o(e_1 \cap e_2 \cap e_3)$. From this we see at once that (12.1h) is satisfied. Moreover, by a similar but easier argument we show that (12.7) also holds.

We can now apply the results of the theory previously developed. First we shall look at the structure of sets which are L-elements or U-elements.

(25.4) **THEOREM.** An element f of F is a U-element if and only if it is an open set; it is an L-element if and only if it is a compact set.

Assume f to be a U-element; let S be a subset of E directed by $\gg$ such that $\bigvee S = f$. By (25.3), $f = \bigvee\{e^o\colon e \text{ in } S\}$, hence is open. Likewise, if f is an L-element, let S be a subset of E directed by $\ll$ such that $\wedge S = F$. Then $f = \wedge\{e^c: e \text{ in } S\}$, and so is compact. Conversely, assume f open. By (8.2d), the set $S = \{e: e \text{ in } E \text{ and } e \ll f\}$ is directed by $\gg$. If p is in f, there is by (25.1ii) an e in E such that p is in e^o and $e^c \subset f^o = f$, so p is in $\vee S$. Thus $f = \vee S$, and is a U-element. Next assume f compact. If p^* is in $T - f$, to each p in f corresponds an e in E such that p is in e^o and $e^c \subset T - \{p^*\}$. The interiors of finitely many such elements $e_1, \ldots, e_n$ of E cover f; their union is an element e of E, and $e \gg f$, and p^* is not in e. By (8.2d), the set $S = \{e: e \text{ in } E \text{ and } e \gg f\}$ is directed by $\ll$, and the arbitrary point p^* of $T - f$ is not in $\wedge S$; hence $f = \wedge S$, and f is an L-element.

If u is a U-element (that is, an open set), it is summable if and only if there exists $S \subset E$ directed by $\gg$ for which $I_0(S)$ is bounded. But then $\vee I_0(S)$ has the same value for all such sets S, in particular for the set $\{e: e \text{ in } E \text{ and } e \ll u\}$. Thus $I_1(u)$ is the supremum of $I_0(e)$ for all sets e of E for which $e \ll u$. This is the same as $I(u)$. If l is an L-element (that is, a compact set), it is necessarily summable, and $I_1(l)$ is the infimum of $I_0(e)$ for all sets e of E for which $e \gg l$. This is the same as $I(l)$.

The symbols $I, \bar{I}, \underline{I}$ will be replaced by m, m^*, m_* respectively (read measure, exterior measure, interior measure respectively). According to the foregoing, the measure of an open set is the supremum of the elementary measures $I_0(e)$ of all sets e of E whose closures are in the open set, provided this supremum is finite; the measure of a compact set is the infimum of the elementary measures $I_0(e)$ of all sets e of E in whose interiors the compact set lies; the exterior measure of a set f is the infimum of mu for all open sets u of finite measure containing f if such sets exist; the interior measure of f is the supremum of ml for all compact sets l contained in f. If m^*f and m_*f are equal, their common value (necessarily finite) is denoted by mf, and f is said to have finite measure.

If f_1 and f_2 are sets of finite measure, by (13.1) and (13.2) so are $f_1 \cup f_2$ and $f_1 \cap f_2$, and $m(f_1 \cup f_2) + m(f_1 \cap f_2) = mf_1 + mf_2$. It is easy to show that the empty set has finite measure, namely measure 0. So if f_1 and f_2 are disjoint, the above equation yields

$$m(f_1 \cup f_2) = mf_1 + mf_2 .$$

Condition (14.1) obviously holds. By (14.3), if $f_1, f_2, f_3, \ldots$ is a sequence of sets of finite measure such that $f_1 \subset f_2 \subset f_3 \subset \ldots$, the union $\bigcup_n f_n$ is a set of finite measure if and only if the numbers

$mf_1, mf_2, \ldots$ are bounded, in which case $m(\bigcup_n f_n) = \bigvee_i mf_i = \lim_{i \to \infty} mf_i$. If $f_1 \supset f_2 \supset f_3 \supset \ldots$, the numbers mf_i are necessarily bounded, so $\bigcap_n f_n$ is of finite measure, and $m(\bigcap_n f_n) = \lim_{i \to \infty} mf_i$. Also, if $(f_\alpha : \alpha \text{ in } A)$ is a directed system of sets of finite measure, all contained in a set f'' of finite measure and σo^*-convergent to a set f_0, by (15.1) f_0 has finite measure, and $mf_0 = \lim_{\alpha \text{ in} A} mf_\alpha$. (We choose f' to be the empty set in hypothesis (ii) of (15.1)).

By (17.1), a set f is measurable if for all pairs f_1, f_2 of sets of finite measure with $f_1 \subset f_2$, the set $\operatorname{mid}(f_1, f_2, f) = f_1 \cup (f_2 \cap f)$ has finite measure. This clearly implies that $f_2 \cap f$ has finite measure, since we can choose f_1 to be the empty set. Conversely, if $f_2 \cap f$ has finite measure, so does $f_1 \cup (f_2 \cap f)$ whenever f_1 has finite measure. So a necessary and sufficient condition that f be measurable is that $f_2 \cap f$ have finite measure whenever f_2 has finite measure. In particular, by (17.5) all open sets and all compact sets are measurable.

We now stop to prove a lemma.

(25.5) **LEMMA**. Let f be a set of finite measure and let g be a subset of f. Then $m_*(f - g) = mf - m^*g$ and $m^*(f - g) = mf - m_*g$.

Let ϵ be a positive number. There exists an open set $u \supset g$ and a compact set $l \subset f$ such that $mu < m^*g + \epsilon$ and $ml > mf - \epsilon$. Then $l - u$ is a compact subset of $f - g$, and $l \cap u$ has finite measure, so

$$m_*(f - g) \geqq m(l - u) = m(l - l \cap u) = ml - m(l \cap u) \geqq ml - mu >$$
$$mf - m^*g - 2\epsilon .$$

Hence $m_*(f - g) \geqq mf - m^*g$. Next, let u be an open set containing f such that $mu < mf + \epsilon$, and let l' be a compact subset of $f - g$ such that $ml' > m_*(f - g) - \epsilon$. Then $l' \subset u$, and $u - l'$ is open and contains g, and $m^*g \leqq m(u - l') = mu - ml' \leqq mf - m_*(f - g) + 2\epsilon$. Hence $m^*g \leqq mf - m_*(f - g)$. With the previous inequality, this establishes the first conclusion. The second is obtained by merely interchanging g and $f - g$.

As a corollary we have

(25.6) **THEOREM**. If f and g have finite measure, so has $f - g$; and if $g \subset f$, $m(f - g) = mf - mg$.

If $g \subset f$ the preceding lemma shows that $m^*(f - g) =$

$m_*(f - g) = mf - mg$, establishing the second conclusion; if f,g have finite measure so has $f \cap g$, and by the second conclusion so has $f - f \cap g$, which is $f - g$.

From this it follows that if f is measurable, so is its complement $T - f$. For let f_2 have finite measure; then $f_2 \cap (T - f) = f_2 - f_2 \cap f$, which has finite measure, and so $f_2 \cap (T - f)$ has finite measure whenever f_2 has. Now the family of measurable sets contains all compact sets and all open sets, and is closed under complementation and under formation of countable unions and intersections. It therefore contains all Borel sets in T.

Let us extend the definition of m^* by setting $m^*f = \infty$ whenever f is not contained in any open set of finite measure. From (25.2) we readily deduce that $m^*(u_1 \cup u_2) \leqq m^*(u_1) + m^*(u_2)$ whenever u_1 and u_2 are open sets; hence $m^*(f_1 \cup f_2) \leqq m^*f_1 + m^*f_2$ for all f_1 and f_2 in F_1. We wish now to show that our definition of measurability is equivalent to Carathéodory's; a set f is measurable if for all sets f', the equation

$$m^*(f' \cap f) + m^*(f' - f) = m^*f' \tag{25.7}$$

is satisfied. This equation is equivalent to

$$m^*(f' \cap f) + m^*(f' - f) \leqq m^*f' \tag{A}$$

since the reverse inequality always holds. (A) is evident if $m^*f' = \infty$, so we assume m^*f' finite. Let $u_1, u_2 \ldots$ be open sets containing f' such that $mu_n < m^*f' + 1/n$; let $g = \bigcap_n u_n$. Then g contains f' and has finite measure, namely m^*f'. Suppose now that f is measurable. Then so is $T - f$, and both $g \cap f$ and $g \cap (T - f)$ have finite measure, and the sum of these measures is mg. Hence (A) holds. Conversely, if (25.7) holds for all f', for all sets f' of finite measure we have $m^*(f \cap f') + m^*(f' - f) = mf'$. By (25.5), $m_*(f \cap f') + m^*(f' - f) = mf'$. Hence the interior and exterior measures of $f \cap f'$ are equal, and $f \cap f'$ has finite measure whenever f' has.

As a special case, let T be n-dimensional Euclidean space, and let E be the set of "figures," whereby "figure" we mean a point set which can be represented as the union of finitely many right-closed intervals R: $\{t: a^i < t^i \leqq b^i, i = 1, \ldots, n\}$. Let ϕ be real-valued and right-continuous on T. To each R (with notation just used) we assign as $I_o(R)$ the sum of 2^n terms, each being the product of the value of ϕ at a vertex of R by a factor which is $+1$ or -1 according as the coordinates of the vertex include an even or an odd number of the a^i. For a figure e which is the union of disjoint right-closed intervals

$R_1, \ldots, R_k$ we define $I_0(e)$ to be $I_0(R_1) + \cdots + I_0(R_k)$; by a well-known elementary computation, this is the same for all such partitions of e. We now make the assumption that $I_0(R) \geqq 0$ for all right-closed intervals R. All the conclusions reached in earlier paragraphs of this section are valid. In addition, we can now verify (15.4) as follows. If e_0 is a right-closed interval $\{t : a^i < t^i \leqq b^i, i = 1, \ldots, n\}$, for each positive integer j we define

$$e_j' = \{x : a^i + \tfrac{3}{j} < t^i \leqq b^i + \tfrac{1}{j}, i = 1, \ldots, n\},$$

$$e_j'' = \{x : a^i + \tfrac{2}{j} < t^i \leqq b^i + \tfrac{2}{j}, i = 1, \ldots, n\},$$

$$e_j''' = \{x : a^i + \tfrac{1}{j} < t^i \leqq b^i + \tfrac{3}{j}, i = 1, \ldots, n\}.$$

All these are contained in a sufficiently large interval. We readily verify (15.4i,ii) while (15.4iii) is a direct consequence of the right continuity of g. So by (15.5) e_0 is of finite measure, and its measure me_0 is the same as the elementary measure $I_0(e_0)$. By addition, this is true for all figures, so in this case the measure m is an extension of the elementary measure I_0 on the class of figures.

As another special case, let M be an arbitrary set, and let T be the cartesian product of copies of [0,1], one for each element of M. That is, the points of T are functions on M to [0,1]. In T we use the weak topology; by Tychonoff's theorem, T is compact. By an interval R in T we mean the set of all points p satisfying a finite number of conditions $a_i < p(m_i) \leqq b_i$ $(i = 1, \ldots, h)$, where $m_1, \ldots, m_h$ are points of M and the a_i and b_i are real numbers such that $0 \leqq a_i \leqq b_i \leqq 1$. The elementary measure of such an interval R will be defined to be $I_0(R) = \Pi_{i=1}^{h} [b_i - a_i]$. We define figures as before; all the conclusions remain valid. Furthermore, as in the finite-dimensional case we show that (15.4) holds, so that the measure m is an extension of the elementary measure I_0.

§ 26. FUNCTIONS ON LOCALLY COMPACT SPACES

In developing the theory of the Lebesgue-Stieltjes integral in n-space we could start with step functions, or with continuous functions which vanish outside some compact set, or we could combine these by using step functions to obtain integrals of continuous functions and proceeding thence from the continuous functions. We shall first consider the generalization of the first method. As in the preceding section, T will be assumed to be a locally compact Hausdorff space; but now F will be the set of all extended-real valued functions on T. Again we assume the existence of a family of sets with the properties (25.1); but this family we now call E^*. We also assume a function on E^* to reals with properties (25.2); but this function we call $(\Delta e : e$ in $E^*)$ instead of $(I_0(e) : e$ in $E^*)$. The class E of "elementary functions" will now consist of all elements e of F with the following property. There exist finitely many disjoint sets $r_1, \ldots, r_n$ of E^* and real numbers $c_1, \ldots, c_n$ such that on r_k the function e has the value c_k; on the complement of $\bigcup_k r_k$, e is 0. Correspondingly, we define $I_0(e)$ to be $c_1 \Delta r_1 + \cdots + c_n \Delta r_n$. By (25.2) it is easily shown that this is uniquely determined.

As usual, for f_1 and f_2 in F the relation $f_1 \geqq f_2$ means $f_1(t) \geqq f_2(t)$ for all t in T. We define $f_1 \gg f_2$ to mean that for each t_0 in T there exists a neighborhood $N(t_0)$ such that $\inf \{f_1(t) : t$ in $N(t_0)\} \geqq \sup \{f_2(t) : t$ in $N(t_0)\}$. We easily show that (22.1c) holds. For G we take the real numbers; this is normal. Then (22.1a,c,d,e) hold. Postulates (22.1g,h,i) are verified without much trouble.

To establish the remaining hypothesis, (22.1f), we let S be a subset of E directed by $\ll$ and such that $\bigwedge S \leqq 0$. Choose an e_0 in S. Let P be the set, belonging to the family E^*, on which $e_0 > 0$; this has compact closure. Let ϵ be a positive number. For each t^* in T, $e(t^*) < \epsilon$ for some e in S. Choose e_1 in S such that $e_1 \ll e$; then $e_1(t) < \epsilon$ for all t in some neighborhood of t^*. A finite number of these neighborhoods, say $N(t^*_1), \ldots, N(t^*_k)$, cover P^c. Let $e_1, \ldots, e_k$ be the corresponding members of S, and let e be a member of S such that $e \ll e_j$, $j = 0, \ldots, k$. Then $e(t) < \epsilon$ for all t in P^c, while $e(t) \leqq 0$ on T - P. Hence $I_0(e) \leqq I_0(\epsilon \chi_P) = \epsilon \Delta P$, and $\bigwedge I_0(S) \leqq \epsilon \Delta P$. Since ϵ is arbitrary, (22.1f) if established. Thus all of (22.1) holds, without the restrictions in the bracketed words. So does (14.1). Postulate (15.4) may or may not be satisfied; if it is, I is an extension of I_0.

By virtue of our definition of $\gg$, it can be shown without difficulty that a function f is a U-function if and only if it satisfies the two conditions (i) there is an elementary function e such that $e \leqq f$, and (ii) f is lower semi-continuous. The dual statement applies to L-functions. If f assumes only 0 and 1 as values, it is the characteristic function of a set; this set is open if and only if f is a U-function, and is compact if and only if f is an L-function. If f is a characteristic function, its upper integral is the same as the infimum of $J_1(u)$ for all summable U-functions u which are characteristic functions of sets; and analogously for the lower integral. Hence the measure derived from this integral I is the same as the measure defined in the preceding section.

Now we make a fresh start. F, F_+ and G shall be as before; but $\gg$ shall now mean merely $\geqq$. Instead of the set E above, we shall use the set E' consisting of all functions ϕ real-valued and continuous on T such that the set $(t : t \text{ in } T \text{ and } \phi(t) \neq 0)$ is contained in a compact set. (The closure of the set on which $\phi \neq 0$ is called the nucleus of ϕ.) Since E' is a sublattice of F and $\geqq$ is reflexive, (22.1g,h) hold. We assume that (22.1e) and the statement about I_0 in (22.1i) hold. The proof of (22.1f) is practically the same as in the first part of this section; instead of χ_P we use a function of E' which is positive on $\bar{P}$ and non-negative everywhere. The other parts of (22.1) are evident. So now define our integral I, and all the theorems of the paper are valid for it. This time (15.4) holds, by (15.6), so I is an extension of I_0.

It is evident that a U-function is lower semi-continuous and exceeds some function of E'. Conversely, if f has these properties, let $f \geqq \phi$, with ϕ in E', and let t' be in T. If $k < f(t')$, there is a neighborhood N of t' with compact closure on which $f(t)$ remains $> k$. Let h be the smaller of k and $\inf \phi$. Construct ψ with $\psi(t') = k$, $\psi(t) = h$ outside N, and $h \leqq \psi \leqq k$ everywhere. Then $\phi \vee \psi$ is in E' and exceeds k at t', so f is the supremum of the set $(e : e \text{ in } E \text{ and } e \leqq f)$, and is a U-function. By similar methods, we can show that the characteristic function of a set M is a U-function if and only if M is open, and is an L-function if and only if M is compact. By (17.5), all open sets and all compact sets are measurable; in particular, T is measurable. If $E_1, E_2, E_3, \ldots$ are measurable sets, for notational simplicity we denote their characteristic functions by $f_1, f_2, f_3, \ldots$ respectively. The characteristic function of $E_1 \cap E_2$ is $f_1 \wedge f_2$, which is measurable by (17.8). The characteristic function of $E_2 - E_1$ is $f_2 - f_1 \wedge f_2$, which is measurable by (21.1) and (21.4). In particular, since E_2 could be chosen to be the whole space, the complement of a measurable set is measurable. The sets $\bigcup_{i=1}^{\infty} E_i$ and $\bigcap_{i=1}^{\infty} E_i$ have the characteristic functions $\lim\sup_i f_i$ and $\lim\inf_i f_i$, which are measurable by (17.8). Hence we see that all Borel sets are measurable. (This discussion of course applies equally well to the first example in this section.)

As a consequence of this, with (23.3), we see that all Borel-measurable functions (in particular, all functions of Baire) are measurable.

Finally, let us return to the first system considered in this section, but use the elementary functions only in order to obtain the integrals $I_1(\phi)$ for continuous functions ϕ with compact nucleus. Since

(22.1e,i) hold with this mapping, we are able to follow from this point on the development in the second example. We shall now show that this gives nothing new; it is merely another path to the same integral I that was defined in the first example. To begin with, we observe from the characterizations of the U-functions in the two examples that these classes are in fact identical, and likewise for the L-functions. Let u be a U-function, and let S' be a subset of E' directed by $\geqq$ and having $\bigvee S' = u$. By (10.7), u is a summable U-element (in the sense of the first method) if and only if the $I_1(S')$ are bounded above, that is, if and only if u is summable in the sense of the second method; and in this case, $\bigvee I_1(S') = I_1(u)$, so the two methods give the same integrals to U-functions, and by duality to L-functions also. It follows at once that the concept of summability and the value of I are the same whichever of the two methods we use.

§ 27. A NON-ABSOLUTELY CONVERGENT INTEGRAL

We shall now construct another example, the resulting integral being in many respects similar to the Perron integral, and perhaps being identical with it. The set F will consist of all extended-real-valued functions on an interval $T = [a,b]$ of real numbers. The relation $\gg$ will be the same as $\geqq$. G will consist of all finitely-additive functions of subintervals of T which depend only on the end-points of the subinterval; $g_1 \leqq g_2$ will mean that $g_1(J) \leqq g_2(J)$ for all subintervals J of T. The class E will consist of all functions e on T with the two properties (i) e is finite-valued, and is continuous at all except countably many points of T; (ii) there exists a continuous finitely-additive function of subintervals of T whose derivative exists and is equal to e at all except countably many points of T. We shall now show that the function of intervals in (ii) is uniquely determined by e; then we shall define $I_0(e)$ to be that function of intervals.

(27.1) **LEMMA**. Let f be a real-valued function on T with the properties:

(a) for $a < x \leqq b$, $f(x-) \geqq f(x)$; and for $a \leqq x < b$, $f(x) \geqq f(x+)$;

(b) except for at most countably many points of T, the lower left derivate of f is $\leqq 0$.

Then f is monotonic non-increasing.

Assume this false. For some c and $d > c$ in T we have $f(d) > f(c)$. If k is a sufficiently small positive number, the same is true of the function g defined by $g(t) = f(t) - kt$. Also, g satisfies hypothesis (a). For each y in the open interval $(g(c),g(d))$, let t_y be the greatest lower bound of points t of $[c,d]$ such that $g(t) \geqq y$. Since by (a) g exceeds y on a neighborhood of d and is less than y on a neighborhood of c, t_y is interior to $[c,d]$. By definition of t_y, there are points t in $[t_y,d]$ arbitrarily close to t_y for which $g(t) \geqq y$; so by (a), $g(t_y) \geqq g(t_y+) \geqq y$. Since $g(t) < y$ for $c \leqq t < t_y$, we also have $g(t_y) \leqq g(t_y-) \leqq y$. Hence $g(t_y) = y$. For t in $[c,t_y)$ the inequality $g(t) < g(t_y)$ holds, and so the lower left derivate of g at t_y is non-negative, and the lower left derivate of f at t_y (which exceeds that of g by k) is positive at t_y. But there are uncountably many points t_y corresponding to y in $(g(c), g(d))$, and we have contradicted hypothesis (b).

To apply this, suppose that g_1 and g_2 correspond to elements e_1 and e_2 of E with $e_1 \geqq e_2$. Define $f(t) = g_2([a,t]) - g_1([a,t])$ for all t in T. Then f is continuous, and its derivative is $\leqq 0$ except at countably many points, so by (27.1) f is non-increasing. Hence $g_2 \leqq g_1$. If $e_2 = e_1$, the reverse inequality also holds.

Now postulates (22.1a,b,c) are obviously satisfied. G is normal, since for each subinterval J of T the function $r(g) = g(J)$ is in R_G, and these functions determine the partial ordering of G; the Dedekind completeness of G is easily verified, so (22.1d) holds. In the preceding paragraph we established (22.1e).

We establish (22.1f) only in the weaker form in which the word "countable" is included. Let $e_1, e_2, e_3 \ldots$ be the members of a set S contained in E, directed by $\leqq$, and having infimum $\leqq 0$. As in (2.4), without loss of generality we may assume $e_1 \geqq e_2 \geqq e_3 \geqq \cdots$. To each e_n corresponds $g_n = I_o(e_n)$. We shall show that (22.1fσ) holds by showing that for all subintervals $[c,d]$ of T, $\lim g_n([c,d]) \leqq 0$. Suppose this false for some $[c,d]$. We define $f_n(t) = g_n([c,t])$ for all t in $[c,d]$; then $f_n(d)$ has a positive lower bound. Let $f_o(t)$ be the limit of the monotone non-increasing sequence $(f_n(t) : n = 1, 2, \ldots)$ for each t in $[c,d]$. Since the e_n are descending, for each subinterval $[r,s]$ of $[c,d]$ and each pair m and $n > m$ of positive integers we have $f_n(s) - f_n(r) \leqq f_m(s) - f_m(r)$, whence $f_o(s) - f_o(r) \leqq f_m(s) - f_m(r)$. If we take $s = d$, we find that f_o satisfies hypothesis (a) of (27.1). Furthermore, if we divide by $s - r$ and let r tend to s, we find that the upper and lower left derivates of f_o at s are at most equal to those of f_m. Thus with the exception of countably many points of $[c,d]$ the lower left derivate of f_o is $\leqq e_m$ for all m, and since e_m approaches a non-positive limit we find that the lower left derivate of f_o is non-positive except at countably many points in $[c,d]$. But by (27.1) this implies that $f_o(d) \leqq f_o(c)$; that is, $\lim g_n([c,d]) \leqq 0$, in contradiction with the assumption concerning the interval $[c,d]$. Hence (22.1fσ) holds.

To prove that (22.1g,h) hold we establish two lemmas.

(27.2) LEMMA. If e is in E and $I_0(e)$ is of bounded variation, $|e|$ is also in E, and $I_0(|e|)$ is the function of intervals whose value for each subinterval T' of T is the total variation of $I_0(e)$ over T'.

Let g be $I_0(e)$, and let $f(t) = g([a,t])$ for all t in T. The total variation $v(t)$ of f on $[a,t]$ is the same as that of g, which by hypothesis is finite; since f is continuous so is v. Except on a countable subset D of T, $e(t)$ is continuous, and $f'(t)$ exists and is equal to $e(t)$. Let t_0 belong to $T - D$. For each positive ϵ there is a neighborhood N of t_0 on which $e(t)$ differs from $e(t_0)$ by less than ϵ. On N the functions whose values are $[e(t_0) + \epsilon](t - t_0) - f(t)$ and $f(t) - [e(t_0) - \epsilon](t - t_0)$ are continuous, and on $N - D$ their derivatives are $e(t_0) + \epsilon - e(t)$ and $e(t) - e(t_0) + \epsilon$, which are non-negative. By (27.1) these functions are non-decreasing on N. So if $[c,d]$ is contained in N, $|f(d) - f(c) - e(t_0)(d - c)| \leqq \epsilon(d - c)$. If we subdivide $[c,d]$ into finitely many subintervals and apply this inequality to each one, we find that the sum of the absolute changes of f over the subintervals differs from $|e(t_0)|(d - c)$ by at most $\epsilon(d - c)$, so that $|v(d) - v(c) - |e(t_0)|(d - c)| \leqq \epsilon(d - c)$. If t is in N, and we apply the preceding inequality with d the larger of t_0 and t and with c the smaller of them, we obtain

$$\left| \frac{v(t) - v(t_0)}{t - t_0} - |e(t_0)| \right| \leqq \epsilon .$$

Therefore $v'(t_0)$ exists and is equal to $|e(t_0)|$. If we now define $g^*(J) = v(d) - v(c)$ for all intervals $[c,d]$ or $(c,d]$ or $[c,d)$ or (c,d) contained in T, we see that g^* has the properties required of $I_0(|e|)$, and the proof is complete.

(27.3) LEMMA. If e_1, e_2 and e_3 are members of E such that $I_0(e_1)$ and $I_0(e_2)$ have a common upper or lower bound in G, then $e_1 \vee e_2$, $e_1 \wedge e_2$ and mid (e_1, e_2, e_3) are also members of E.

By (6.7), $I_0(e_1)$ and $I_0(e_2)$ have both an upper bound and a lower bound in G. Let g_i be $I_o(e_i)$, $i = 1, 2$, and let g_3 be an upper bound for g_1 and g_2. Then $g_3 - g_1$ and $g_3 - g_2$ are non-negative, hence of bounded variation; so is their difference $g_2 - g_1$. But $g_2 - g_1 = I_o(e_2 - e_1)$, so $|e_2 - e_1|$ is in E. So are $e_1 \vee e_2 = (e_1 + e_2 + |e_2 - e_1|)/2$ and $e_1 \wedge e_2 = (e_1 + e_2 - |e_2 - e_1|)/2$.

Since mid (e_1,e_2,e_3) = mid $(e_1 \vee e_2, e_1 \wedge e_2, e_3)$ by (2.13), and $e_1 \vee e_2$ and $e_1 \wedge e_2$ are in E, we may simplify the notation by assuming that $e_1 \leqq e_2$. Also, since mid (e_1,e_2,e_3) = mid $(0, e_2 - e_1, e_3 - e_1) + e_1$, it is sufficient to prove that the first term in the right member belongs to E; that is, we need only prove that mid $(0,e_2,e_3)$ is in E when e_2 and e_3 are in E and $e_2 \geqq 0$. By (2.12), this implies mid $(0,e_2,e_3) = 0 \vee (e_2 \wedge e_3) = e_2 \wedge (e_3 \vee 0)$.

If e_2 is bounded, mid $(0,e_2,e_3)$ is bounded, and being continuous at all but countably many points it is Riemann integrable. The function of intervals

$$g([c,d]) = \int_c^d \text{mid } (0,e_2,e_3)\, dt$$

serves as $I_o(\text{mid } (0,e_2,e_3))$. If e_2 is not bounded, let $e_n = e_2 \wedge n$, $n = 4, 5, 6, \ldots$, and let

$$f_n(t) = \int_a^t \text{mid } (0,e_n,e_3)\, d\tau .$$

The differences $f_{n+1} - f_n$ are monotonic non-decreasing, and for all n we have

$$f_n(t) \leqq \int_a^t e_n\, d\tau \leqq g_2([a,b]) ,$$

so the f_n have a limit f_o. Except on a countable set D, e_2 and e_3 are continuous; each point t_o of T - D is contained in a neighborhood on which e_2 is bounded, so $e_2 = e_n$ on this neighborhood for all large n. For all such n the difference $f_n - f_o$ is constant on this neighborhood, so

$$f'_o = f'_n = \text{mid } (0,e_n,e_3) = \text{mid } (0,e_2,e_3) .$$

Hence the function of intervals defined by $g([c,d]) = f_o(d) - f_o(e)$ serves as $I_o(\text{mid } (0,e_2,e_3))$, and the proof is complete.

From (27.3) we see at once that (22.1g,h) are satisfied; and (22.1i) is obvious. So the postulates (22.1σ) are verified. Postulate (14.1) is also satisfied. For let S_1, S_2, ... be subsets of G directed by $\leqq$ and having $\bigwedge S_n = 0$. For each n, select from S_n a member g_n such that $g_n(T) < 2^{-n}$. Then for each subinterval $[c,d]$ of T we have $0 \leqq g_n([c,d]) \leqq g_n(T) < 2^{-n}$, so Σg_n exists in G, and is an upper bound for all partial sums $g_1 + \cdots + g_k$.

Let us denote by E' the set of all functions continuous on T; for e in E', $I_0(e)$ is the function of intervals whose value for each $[c,d] \subset T$ is the Cauchy integral of e over $[c,d]$. Let $I_0'(e)$ be the value of $I_0(e)$ corresponding to the whole interval T. If we start with E' and I_0', the resulting integral is the Lebesgue integral, as we developed it in the preceding section. Here it is immaterial whether or not we restrict the sets S used in defining U- and L-functions to be countable, since we obtain the same suprema if we use only the polynomials with rational coefficients. If u is a summable U-function as developed from E' and I_0', it is also a summable U-function in the sense of this section. Also, $I_1'(u)$ is the same as the value of $I_1(u)$ corresponding to T. A like statement holds for L-functions. It follows readily that if f is Lebesgue integrable, it is summable in the sense of this section, and its Lebesgue integral is the value of $I(f)$ corresponding to T. But E' is directed by $\geqq$, so by (16.2 vi) if f is Lebesgue integrable it belongs to the class F_0 containing the zero-function. Hence the integral defined in this section is a generalization of the Lebesgue integral, the Lebesgue-summable functions all being contained in the class F_0. We now sketch the proof that the Lebesgue summable functions in fact constitute F_0.

If e is in E and is bounded, it is Riemann integrable, being continuous except on a countable set. If e is in E and is bounded below, for each positive integer n the function $e \cap n$ is Riemann integrable, and the integrals rise boundedly with n, so $e = \lim e \wedge n$ is Lebesgue summable. If u is a non-negative summable U-function and e is in E and $e \leqq u$, then $I_0(e)$ and $I_0(0)$ have a common upper bound in G, so $e \vee 0$ is in E. Thus u may be considered to be the limit of a rising sequence of non-negative members of E; hence if u is summable in the sense of this section it is Lebesgue summable. A similar statement applies to non-negative L-functions. If f is non-negative and summable in the present sense, we can find sequences u_i, l_i $(i = 1, 2, \ldots)$ of non-negative U- and L-functions such that $l_i \leqq f \leqq u_i$ and $\lim_i I_1(u_i) = \lim I_1(l_i) = I(f)$ for the interval T. All the u_i and l_i are Lebesgue summable, and their Lebesgue integrals are the same as the $I_1(u_i)$ and $I_1(l_i)$ corresponding to T, so $\lim u_i$ and $\lim l_i$ are Lebesgue summable, and have the same integral. Since f is between them, it is too Lebesgue summable. Finally, if f is in F_0 both $f \vee 0$ and $-(f \wedge 0)$ are summable and non-negative, hence Lebesgue summable; so f also is Lebesgue summable.

There are various well-known extensions of the Riemann integral to unbounded functions, some of which are non-absolutely convergent. All these extensions apply to integrands whose infinite discontinuities form a countable closed set D; the integrals are continuous functions, and on each interval [c,d] contained in T and containing no point of D the integral is the Riemann integral. Such integrals belong to our class E, so these extensions of the Riemann integral are special cases of the integral here defined.

It is easy to show (cf. [7], p. 197) that if u is a summable U-function, the lower derivative of $I_1(u)$ is everywhere at least equal to U. Hence if f is in F and u is a summable U-function $\geqq$ f, u is also a Perron major function for f. Likewise, if l is a summable l-function $\leqq$ f, it is a Perron minor function for f. Hence the integral of this section is not more general than the Perron integral. Whether it is the same as the Perron integral I am unable to state.

§ 28. REMARKS ON OPERATORS IN HILBERT SPACES

In the next section we shall discuss the spectral resolution of a bounded hermitian operator on a Hilbert space H— that is, a complete unitary space, not necessarily separable. In preparation, we now establish a few properties of such operators, most of these properties being well known.

Let G be the set of all bounded hermitian operators on H. Each bounded linear operator B on H determines a function ((Bx,x) : x in H), and this function is real-valued if B is hermitian. As usual, we define B $\geqq$ 0 to mean that (Bx,x) $\geqq$ 0 for all x in H; and we define B" $\geqq$ B' to mean B" - B' $\geqq$ 0. This relation $\geqq$ clearly satisfies (1.1a). To show that it satisfies (1.1b), (that is, is proper) we make a simple calculation:

(28.1) LEMMA. If B is a bounded linear operation on H, and x and y are in H, then

$$2(Bx,y) = (B[x + y],[x + y]) + i(B[x + iy],[x + iy]) - (1 + i)\,[(Bx,x) + (By,y)]\,.$$

Now if B" $\geqq$ B' and B' $\geqq$ B", then the difference B = B" - B' satisfies (Bx,x) = 0 for all x in H, so by (28.1) we have (Bx,y) = 0 for all x and y in H, whence B = 0. Thus $\geqq$ is a (proper) partial ordering.

(28.2) LEMMA. Let S be a subset of G directed by $\geqq$

and having an upper bound B' in G. Then there exists a bounded hermitian operator C such that

(i) $(Cx,x) = \bigvee\{(Bx,x) : B \text{ in } S\}$ for each x in H;

(ii) $\lim_{B \text{ in } S} (Bx,y) = (Cx,y)$ for each x and y in H;

(iii) $C = \bigvee S$.

For each x in H, $((Bx,x) : B \text{ in } S)$ is an isotone net of real numbers, so has a limit which is its supremum. Each term in the right member of the equation in (28.1) has a limit, so for each x and y in H the net (Bx,y) also has a limit. If B'' is in S, eventually $B'' \leqq B \leqq B'$. Let M be the larger of $\|B'\|$, $\|B''\|$. For all elements B_0 of G, $\|B_0\|$ is the supremum of $|(B_0x,x)|$ for all unit vectors x in H. For such x, eventually $-M \leqq (B''x,x) \leqq (Bx,x) \leqq (B'x,x) \leqq M$, whence $\|B\| \leqq M$. Thus for all x,y in H we eventually have $|(Bx,y)| \leqq M\cdot|x|\cdot|y|$, and the limit of (Bx,y) is bounded for x,y in the unit sphere. Since for each fixed x the function (y,Bx) is linear in y, the same is true of its limit; being bounded on the unit sphere, this limit is for each x the inner product of y with some element of H, which we shall call Cx. Thus $\lim (Bx,y) = (Cx,y)$ for all x and y in H. If in particular we choose $y = Cx$, the previous estimate shows that $|Cx| \leqq M\cdot|x|$. Also, for each fixed y the function (Bx,y) is linear in x, so (Cx,y) is also linear in x. Thus $(C[k'x' + k''x''] - k'Cx' - k''Cx'', y)$ vanishes for all elements x',x'',y of H and all complex numbers k',k'', whence C is a linear transformation. We have already shown it bounded. Also, for all a and y in H,

$$(Cx,y) = \lim (Bx,y) = \lim (x,By)$$

$$= \lim \overline{(By,x)} = \overline{(Cy,x)} = (x,Cy) \;,$$

so C is hermitian. This completes the proof of conclusion (ii). Conclusion (i) follows at once, by the second remark after (3.1).

By (i), C is an upper bound for S in G. Let C' be also an upper bound for S in G. Then for all x in H we have $(C'x,x) \geqq (Bx,x)$ for all B in S, so by (i) $(C'x,x) \geqq (Cx,x)$, and $C' \geqq C$. This establishes (iii) and completes the proof.

This lemma allows us to prove that in G, o-convergence is equivalent to eventually-bounded strong convergence:

(28.3) LEMMA. If $(B_\alpha : \alpha \text{ in } A)$ is a net in G and

B_0 is in G, then o-lim $B = B_0$ if and only if the norms $\|B_\alpha\|$ are eventually bounded and $\lim |B_\alpha x - B_0 x| = 0$ for every x in H.

Assume o-lim $B_\alpha = B_0$; we shall give the proof, due to J. M. G. Fell and J. L. Kelley (Fell and Kelley [1]), that B_α tends strongly to B_0. Let M,N be sets as in (3.1); without loss of generality we may assume that all their elements have norms less than some positive finite number b. Let x and y be unit vectors in H and ϵ a positive number. By (28.2) we can find m in M and n in N such that

$$([B_0 - m]x,x) < \epsilon^2/16b \ , \quad ([n - B_0]x,x) < \epsilon^2/16b \ ,$$

whence

$$([n - m]x,x) < \epsilon^2/8b \ .$$

Eventually, $m \leqq B_\alpha \leqq n$; then $\|B_\alpha\| \leqq \max\{\|m\|,\|n\|\}$, and $B_\alpha - m$ is semi-definite, and the Cauchy-Schwarz inequality can be applied:

$$\begin{aligned} |([B_\alpha - m]x,y)|^2 &\leqq ([B_\alpha - m]x,x) \cdot ([B_\alpha - m]y,y) \\ &\leqq ([n - m]x,x) \cdot ([n - m]y,y) \\ &< (\epsilon^2/8b) \cdot 2b = \epsilon^2/4 \ . \end{aligned}$$

This is valid for all unit vectors y, so

$$|[B_\alpha - m]x| < \epsilon/2 \ .$$

The same reasoning applies with B_0 in place of B_α, so $|[B_0 - m]x| < \epsilon/2$, and thus eventually $|[B_\alpha - B_0]x| < \epsilon$. Since now $\lim |[B_\alpha - B_0]x| = 0$ for all unit x, the same holds for all x, and B_α converges strongly to B_0.

Conversely, assume that B_α converges strongly to B_0 and the norms $\|B_\alpha\|$ are eventually bounded. Let $(\phi_\gamma : \gamma \text{ in } \Gamma)$ be a complete orthonormal set in H, and let b be an upper bound for the norms of the operators $B_\alpha - B_0$ for all $\alpha \geqq$ a certain α_0. To each finite subset $\sigma = [\gamma_1, \ldots, \gamma_{s(\sigma)}]$ of Γ corresponds a projection P_σ on the space generated by the ϕ_γ with γ in σ; thus $P_\sigma x = \Sigma_{\gamma \text{ in } \sigma} (x,\phi_\gamma)\phi_\gamma$. Let n_σ be the operator defined by $n_\sigma x = x/s(\sigma) + b(1 - P_\sigma)x$; this is hermitian, and the set of all n_σ is directed by $\leqq$ and has 0 for infimum. If we define N to be the set of all operators $B_0 + n_\sigma$ and M

to be the set of all operators $B_0 - n_\sigma$, postulates (3.1a,b) are satisfied, and $\bigvee M = B_0$. It remains to prove that if m is in M and n in N we eventually have $m \leqq B_\alpha \leqq n$. It will be enough to show that for each finite set $\sigma \subset \Gamma$, we eventually have $B_\alpha - B_0 \leqq n_\sigma$. If σ consists of $\gamma_1, \ldots, \gamma_{s(\sigma)}$, we eventually have

$$|[B_\alpha - B_0]\phi_\gamma| < \tfrac{1}{2}\, s(\sigma)^2 \quad (\gamma = \gamma_1, \ldots, \gamma_{s(\sigma)})$$

Then

$$([B_\alpha - B_0]x,x) = ([B_\alpha - B_0]P_\sigma x,x) + ([B_\alpha - B](1 - P_\sigma)x,P_\sigma x)$$
$$+ ([B_\alpha - B_0](1 - P_\sigma)x,(1 - P_\sigma)x) .$$

For the first term we have

$$\begin{aligned} |([B_\alpha - B_0]P_\sigma x,x)| &\leqq |[B_\alpha - B_0]P_\sigma x| \\ &\leqq \textstyle\sum_{\gamma \text{ in } \sigma} |(x,\phi_\gamma)|\cdot|[B_\alpha - B_0]\phi_\gamma| \\ &\leqq \tfrac{1}{2}\, s(\sigma) . \end{aligned}$$

For the second term, written in the form $((1 - P_\sigma)x,[B_\alpha - B_0]P_\sigma x)$, we have the same estimate, since $|(1 - P_\sigma)x| \leqq 1$. For $\alpha \geqq \alpha_0$ the last term is at most $b|(1 - P_\sigma)x|^2$. Hence for such α

$$|([B_\alpha - B_0]x,x)| \leqq \frac{1}{s(\sigma)} + b|(1 - P_\sigma)x|^2 .$$

But by definition

$$\begin{aligned} (n_\sigma x,x) &= \frac{(x,x)}{s(\sigma)} + b([1 - P_\sigma]x,x) \\ &= \frac{1}{s(\sigma)} + b|(1 - P_\sigma)x|^2 \end{aligned}$$

So when $\alpha \geqq \alpha_0$, $([B_\alpha - B_0]x,x) \leqq (n_\sigma x,x)$ for all unit vectors x, hence for all x, and the proof is complete.

(28.4) LEMMA. The set G of bounded hermitian operators is Dedekind complete.

This is an immediate consequence of (28.2) and its dual.

(28.5) LEMMA. The set G is normal.

Each x in H provides a real-valued function r on G, defined by $r(B) = (Bx,x)$, B in G. Let S',S'' be subsets of G directed by $\geqq, \leqq$ respectively, such that the quantities $C' = \bigvee S'$ and $C'' = \bigwedge S''$ exist. If $\bigvee S' \geqq \bigwedge S''$, then $C' - C'' \geqq 0$, and by (28.2)

$$\lim_{B' \text{ in } S', B'' \text{ in } S''} ([B' - B'']x,x) = ([C' - C'']x,x) \geqq 0$$

for all x. That is, $\bigvee r(S') \geqq \bigwedge r(S'')$, and r is smoothly rising. If $r(B'') \geqq r(B')$ for all these r, it follows immediately from the definition that $B'' \geqq B'$.

(28.6) **LEMMA**. If A and B are bounded hermitian operators on H, and $A \geqq 0$ and $B \geqq 0$, and $AB = BA$, then $AB \geqq 0$.

The proof, due to F. Riesz, we take almost verbatim from Nagy (Nagy[1]). If $A = 0$ the conclusion is trivial, so we assume $||A|| > 0$. Define $A_1 = (||A||)^{-1}A$, $A_1 = A_1 - A_1^2$, ..., $A_{n+1} = A_n - A_n^2$, ... By induction we establish

$$(*) \qquad 0 \leqq A_n \leqq 1, \quad n = 1, 2, \ldots .$$

This is obvious for $n = 1$. Assume it true for $n = k$. The definition of A_{k+1} can be re-written as

$$A_{k+1} = A_k^2(1 - A_k) + A_k(1 - A_k)^2 ,$$

$$1 - A_{k+1} = (1 - A_k) + A_k^2 .$$

Both terms in the right member of the last equation are $\geqq 0$, so $1 - A_{k+1} \geqq 0$. By the first equation, for each x in H we have

$$(A_{k+1}x,x) = ([1 - A_k]A_kx,A_kx) + (A_k[1 - A_k]x, [1 - A_k]x)$$
$$\geqq 0 .$$

So $A_{k+1} \geqq 0$, and (*) is established.

By combining the definitions of the A_k,

$$A_1 = A_1^2 + A_2^2 + \ldots + A_n^2 + A_{n+1} ,$$

whence for each x in H

$$\sum_{k=1}^{n} ||A_k x||^2 = \sum_{k=1}^{n} (A_k x, A_k x)$$

$$= \sum_{k=1}^{n} (A_k^2 x, x)$$

$$= (A_1 x, x) - (A_{n+1} x, x)$$

$$\leqq (A_1 x, x) ,$$

since $A_{n+1} \geqq 0$. Hence $\Sigma ||A_k x||^2$ is convergent, and therefore $\lim ||A_k x|| = 0$. Now

$$\lim_{n \to \infty} || \sum_{k=1}^{n} A_k^2 x - Ax || = \lim_{n \to \infty} || A_{n+1} x || = 0 .$$

By hypothesis B commutes with A, hence by induction it commutes with all the A_k. Hence $(BA_k^2 x, x) = (BA_k x, A_k x) \geqq 0$, and so

$$(BAx, x) = ||A||(BA_1 x, x)$$

$$= ||A|| \left\{ \sum_{k=1}^{n} (BA_k^2 x, x) + (BA_{n+1} x, x) \right\}$$

$$\geqq ||A|| \ (BA_{n+1} x, x) .$$

Letting $n \to \infty$ yields $(BAx, x) \geqq 0$, and the proof is complete.

§ 29. SPECTRAL RESOLUTION OF A BOUNDED HERMITIAN OPERATOR

Let B be a bounded hermitian operator on the Hilbert space H, and let T be the interval $[-||B||, ||B||]$. We define F and F_+ as in (22.1a,b). We define E to be the set of all real polynomials on T. For f_1 and f_2 in F, the statement $f_1 \gg f_2$ shall mean that $f_1(t) - f_2(t)$ is defined for all t in T and has a positive lower bound. Then (22.1a,b,c) are satisfied. So is (22.1g), irrespective of the definition of I_o, since E is directed by $\gg$ and by $\ll$.

Let $p = (a_o + a_1 t + \dots + a_n t^n : t \text{ in } T)$ be a member of E. We define $I_o(p)$ to be the operator commonly called $p(B)$, namely

$$I_o(p) = a_o 1 + a_1 B + \dots + a_n B^n .$$

This has domain E and range contained in G. In order to prove it isotone, it is sufficient to show that if $p(t) \geqq 0$ for all t in T, then $I_o(p) \geqq 0$. Assume p non-negative on T, and factor it into linear factors. The non-real roots occur in conjugate pairs. We multiply such pairs; for example, the pair $t - (a + ib), t - (a - ib)$ have as product $(t - a)^2 + b^2$. The other factors consist of powers of distinct linear factors $(t - r_i)^{k_i}$ and the constant factor a_k. An even number of these are negative on T; we change the signs of these, and obtain the factorization

$$p(t) = |a_k| (c_1-t)^{k_1} \ldots (c_h-t)^{k_h}([t-a_1]^2 + b_1^2) \ldots ([t-a_k]^2 + b_k^2) ,$$

where the factors $(c_i - t)^{k_i}$ are non-negative on T. Then

$$p(B) = |a_k|(c_1 1 - B)^{k_1} \ldots (c_h 1 - B)^{k_h}([B - a_1 1]^2 + b_1^2 1) \ldots$$
$$([B - a_k 1]^2 + b_k^2 1) .$$

The quadratic factors are clearly $\geqq 0$. Since $(c_i - t)^{k_i} \geqq 0$ on T, either k_i is even or else we have $c_i \geqq ||B||$, so that $c_i 1 - B \geqq 0$. In either case, $(c_i 1 - B)^{k_i} \geqq 0$. By (28.6), $p(B) \geqq 0$, and thus (22.1e) is established.

The statement about I_o in (22.1i) is an easy consequence of the definition, and the other parts of (22.1i) are obvious. If e_1, e_2 and e_3 are in E, and $f \gg \text{mid}(e_1,e_2,e_3)$, there exists $\epsilon > 0$ such that $f \geqq 2\epsilon + \text{mid}(e_1,e_2,e_3)$. The function $\text{mid}(e_1,e_2,e_3) + \epsilon$ is continuous, and can be approximated uniformly on T to within $\epsilon/2$ by a polynomial e; then this e serves in (22.1h). If S is a subset of E directed by $\ll$ and having $\bigwedge S \leqq 0$ (wherein 0 stands for the function which vanishes identically on T), by an argument already used several times we find that for every positive ϵ there is an e in S such that $e(t) < \epsilon$ for all t in T. Then $\epsilon - e$ is a positive-valued polynomial on T, so $\epsilon 1 - I_o(e)$ is a non-negative hermitian operator, and $I_o(e) \leqq \epsilon 1$. For each x in H, $(I_o(e)x,x) \leqq (\epsilon x,x) = \epsilon |x|^2$, so the infimum of $(I_o(e)x,x)$ for B in S is $\leqq 0$. This establishes (22.1f).

We can also verify that (15.4) holds. For any e_o in E we define $f = e_o - 1$, $g = e_o + 1$; and for all positive integers j we define $e'_j = e_o - 1/j$, $e''_j = e_o$, $e'''_j = e_o + 1/j$; then (15.4) clearly holds. The postulates being satisfied without the restrictions in the brackets, and (15.4) being satisfied, we can extend the domain of definition of I_o and obtain an "integral," or mapping, I, defined on the class of summable functions in the class F. However, in accordance with established usage we shall usually denote the image I(f) of a summable function f in F by the symbol f(B). This symbol has already been defined when f is a polynomial; but by (15.5) there is no contradiction.

From (19.3), (19.4), (11.8) and (13.1) (with (13.3)) we obtain the following theorem.

(29.1) **THEOREM.** Let f_1 and f_2 be summable functions in the class F, and let c_1 and c_2 be real numbers. Then

(i) if f_1 and f_2 are finite-valued and $f = c_1 f_1 + c_2 f_2$, f is also summable, and $f(B) = c_1 f_1(B) + c_2 f_2(B)$;

(ii) if $f_1 \leqq f_2$, then $f_1(B) \leqq f_2(B)$;

(iii) $|f_1|$ is summable, and $-|f_1|(B) \leqq f_1(B) \leqq |f_1|(B)$.

From (15.1) and (3.5) we obtain

(29.2) **THEOREM.** Let $f_o, f_1, f_2, \ldots$ be uniformly bounded functions in the class F such that $\lim_{n \to \infty} f_n(t) = f_o(t)$ for all t in T. If $f_1, f_2, \ldots$ are summable, so is f_o, and $f_o(B) = \text{o-}\lim_{n \to \infty} f_n(B)$.

The class of summable functions includes all polynomials, and by (29.2) is closed under uniformly bounded passage to the limit, hence the following statement holds.

(29.3) **THEOREM.** If f is a bounded function of Baire on T, it is summable.

In order to study products, it is convenient to define a class of operators which has proved itself useful in several connections. The product of bounded hermitian operators A and B is a bounded linear operator, but is hermitian if and only if A and B commute. Let $cc(B)$ denote the class of all bounded hermitian operators which commute with every bounded hermitian operator which commutes with B. Then every two elements of $cc(B)$ commute, and their product is again in $cc(B)$. Also $cc(B)$ is linear, so (7.5) is satisfied. By this and (28.6), (7.6a,b,c) are satisfied. Suppose that A commutes with B, and that $(C_\gamma : \gamma \text{ in } \Gamma)$ and $(D_\delta : \delta \text{ in } \Delta)$ are nets of elements of $cc(B)$ having the respective o-limits C and D. For each x in H and each γ in Γ we have $C_\gamma A x = A C_\gamma x$, so by (28.3) $CAx = ACx$. That is, $cc(B)$ is closed under o-convergence, and by (4.2) is Dedekind closed. Also, for each x in y in H we have

$$\lim_{\gamma \text{ in } \Gamma, \delta \text{ in } \Delta} (D_\delta x, C_\gamma y) = (Dx, Cy) ,$$

whence $\lim_{\gamma,\delta} (C_\gamma D_\delta x, y) = (CDx, y)$. Now suppose the C_γ and D_δ all $\geqq 0$. If the nets $(C_\gamma : \gamma \text{ in } \Gamma)$ and $(D_\delta : \delta \text{ in } \Delta)$ are both isotone, so is the net $(C_\gamma D_\delta : \gamma \text{ in } \Gamma \text{ and } \delta \text{ in } \Delta)$, so by the Dedekind completeness of G it has an o-limit E. Then $\lim_{\gamma,\delta} (C_\gamma D_\delta x, y) = (Ex, y)$ for all x,y in H, whence $E = CD$. A similar proof holds if the nets $(C_\gamma : \gamma \text{ in } \Gamma)$ and $(D_\delta : \delta \text{ in } \Delta)$ are antitone. By (5.5), the product is a smoothly-rising function of pairs of non-negative elements of cc(B), and so (7.6d) is satisfied.

(29.4) **THEOREM.** If f is summable, f(B) is in cc(B).

Obviously if e is in E, e(B) is in cc(B). If u is a summable U-function, let S be a subset of E associated with u. Then $u(B) = I_1(u) = \bigvee I_0(S) = \bigvee\{e(B) : e \text{ in } S\}$, and since cc(B) is Dedekind closed $I_1(u)$ is in cc(B). By (11.9), if f is summable $f(B) = I(f) = \bigwedge I_1(U[\geqq f]) = \bigwedge\{u(B) : u \text{ in } U[\geqq f]\}$, so it too is in cc(B).

(29.5) **THEOREM.** If f_1 and f_2 are summable, so is their product $p = f_1 f_2$, and $p(B) = f_1(B) f_2(B)$.

If e_1 and e_2 are in E, we obtain at once from the definition of I_0 that $I_0(e_1 e_2) = I_0(e_1)\, I_0(e_2)$. Our theorem now follows by (19.5).

This theorem shows that the mapping I provides an operational calculus for functions of B.

If M is a measurable subset of the interval T, its measure is by (22.5) the integral of its characteristic function χ_M. In accordance with the notation we are using, this would be $\chi_M(B)$. We shall also use the symbol $m_B M$ for it. This is certainly defined for all Borel subsets of T. By (29.5), for any two measurable subsets M,N of T we have

$$\chi_M(B)\chi_N(B) = \chi_N(B)\chi_M(B) = \chi_{M\cap N}(B) .$$

In particular, if we take $M = N$ we obtain $[\chi_M(B)]^2 = \chi_M(B)$. But this, together with the fact that $\chi_M(B)$ is a bounded hermitian operator, implies that it is a projection. Hence we have the following theorem.

(29.6) **THEOREM.** Each measurable set M contained in T

has a measure which is a projection operator in H. The product of the measures of two measurable sets M,N, taken in either order, is the measure of their intersection; if $M \subset N$, then $m_B(M) \leqq m_B(N)$, and the range of the projection $m_B M$ is contained in the range of $m_B N$; if M and N are disjoint, the ranges of the projections $m_B M$ and $m_B N$ are orthogonal. (E. R. Lorch seems to have been the first to consider projection-valued measures; see (Lorch [1]).)

This system of measures is in fact the resolution of the identity corresponding to B. In terms of it, by means of the Lebesgue process, we can reconstruct the integral f(B) for any function f summable on the interval T, in particular for any bounded Baire function. However, in the present case we can say more than we did in (23.5). For the functions constant on measurable sets which occur in connection with a Lebesgue ladder differ from the integrand function by at most the norm of the ladder, and so tend uniformly to the integrand; whence we deduce that the Lebesgue sums are bounded hermitian operators which approach the integral of f in the uniform sense, the norm of the difference approaching zero.

Let us agree to say that a measure m is <u>regular</u> if for each set M measurable with respect to m, there are sets A*, B* which are respectively the union of countably many compact sets and the intersection of countably many open sets, such that $A^* \subset M \subset B^*$ and $mA^* = mM = mB^*$.

(29.7) **THEOREM**. For each x in H, the set function μ_x defined for each measurable set M by $\mu_x M = ([m_B M]x,x)$ is a non-negative regular measure.

It is obvious that μ_x is non-negative and countably additive. From the definition, all U-functions are lower semi-continuous and all L-functions are upper semi-continuous; and from the definition, for each measurable set M the value of $([m_B M]x,x)$ is the infimum of $(u(B)x,x)$ for all summable U-functions $u \geqq \chi_M$. Let ϵ be positive; there is a U-function $u \geqq \chi_M$ such that $(u(B)x,x) < ([m_B M]x,x) + \epsilon$. For a number $k > 1$ and close enough to 1 we will have $k(u(B)x,x) < ([m_B M]x,x) + \epsilon$. Let O_ϵ be the set on which $ku > 1$; this is open and contains M, and its characteristic function is at most equal to ku. Hence $\mu_x O_\epsilon \leqq (ku(B)x,x) \leqq ([m_B M]x,x) + \epsilon$. Now we let ϵ take on the values 1, 1/2, 1/3, ... successively, and define the set B* to be the intersection of the corresponding sets O_ϵ. This has the desired structure, contains M and satisfies $\mu_x M = \mu_x B^*$. The set A^* is constructed analogously.

(29.8) COROLLARY. If H is separable, the measure function m_B is regular.

Let $x_1, x_2, \ldots$ be a countable set of points dense in H. For each x_n there is a pair of sets A^*_n, B^*_n with the properties stated in (29.7). Let A^* be the union of the A^*_n and B^* the intersection of the B^*_n. These have the desired structure, and $A^* \subset M \subset B^*$. Obviously $([m_B A^*]x_n, x_n) = ([m_B M]x_n, x_n) = ([m_B B^*]x_n, x_n)$ for all n, whence by the continuity of the operators and the density of the x_n the same holds for all points x of H. Hence $m_B A^* = m_B M = m_B B^*$, and the proof is complete.

The measure defined by our integration is maximal with respect to the regularity property in (29.7). For suppose that m' is an extension of m_B, and let M be a set measurable with respect to m'. Since m' and m_B coincide on Borel sets, and (29.7) holds for m', for each x in H there are sets A^* and B^* such that $([m_B A^*]x, x) \leqq (\underline{I}(\chi_M)x, x) \leqq (\overline{I}(\chi_M)x, x) \leqq ([m_B B^*]x, x)$. So the middle pair of inner products have the same value for each x in H, and the upper and lower integrals of the characteristic function of M are equal; that is, M is measurable. From the preceding relation and the equality of $([m_B A^*]x, x)$ and $([m' A^* x], x)$ we also deduce $([m_B M]x, x) = ([m'M]x, x)$ for all x, so that $m'm = m_B M$, and m' is not a proper extension of m_B. From our measure function we can form the more traditional form of resolution of the identity by defining E_λ to be $m(-\infty, \lambda]$ for all real λ. This has the standard properties (see, e.g., (von Neumann [1])); but for brevity we shall end the discussion here.

§ 30. BOCHNER'S GENERALIZATION OF THE BERNSTEIN-WIDDER THEOREM

In order to state the theorem which we are about to establish, it is convenient to introduce some new terminology. If f is defined on a subset of the reals and its values lie in an additive group, and α and $h (h > 0)$ are numbers such that α and $\alpha + h$ are in the domain of f, we define $\Delta_h^0 f(\alpha) = f(\alpha)$, $\Delta_h^1 f(\alpha) = f(\alpha + h) - f(\alpha)$; and inductively we define $\Delta_h^{n+1} f(\alpha) = \Delta_h^1[\Delta_h^n f(\alpha)]$. If f is defined on the open half-line $\{\alpha: 0 < \alpha < \infty\}$, it is <u>completely monotonic</u> on $\{\alpha: 0 < \alpha < \infty\}$ if its range is partially ordered and the inequalities $(-1)^n \Delta_h^n f(\alpha) \geqq 0$, $n = 0, 1, 2, \ldots$ hold for all $\alpha > 0$; if f is defined on the closed half-line $\{\alpha: 0 \leqq \alpha < \infty\}$, it is <u>completely monotonic</u> on that half-line if the same inequalities hold for $\alpha \geqq 0$.

The generalized form of the Bernstein-Widder theorem which we shall establish is essentially identical with that of Bochner (Bochner [2]):

(30.1) **THEOREM.** Let $g(\alpha)$: $0 < \alpha < \infty$ be a function with values in a Dedekind σ-complete, normal partially-ordered linear space G. If g is completely monotonic, there exists a countably additive non-negative measure function μ, defined on a family of subsets of $(0,\infty)$ which includes all bounded Borel sets and assuming values in G, such that

$$g(\alpha) = \int_0^\infty \exp(-\alpha t)\mu(d\alpha) \; ;$$

and conversely.

We prove only the direct implication; the easy proof of the converse we omit.

In the proof we shall make use of certain properties of the Bernstein polynomials. If f is defined on $[0,1]$, for each positive integer k we define the k-th Bernstein polynomial of f to be the polynomial $B_k f$ defined by

$$B_k f(x) = \sum_{m=0}^{k} f(\tfrac{m}{k})\binom{k}{m} x^m (1-x)^{k-m} .$$

(30.2) **LEMMA.** If f is a polynomial of degree d, there exist polynomials $p_0, p_1, \ldots, p_d$ of degree at most d such that for each integer $k \geqq d$ and each x in $[0,1]$,

$$f(x) - B_k f(x) = \sum_{m=1}^{d} k^{-m} p_m(x) .$$

In the identity

$$(e^y + z)^k = \sum_{m=0}^{k} \binom{k}{m} e^{my} z^{k-m}$$

we differentiate repeatedly with respect to y, obtaining the partial derivatives of orders $\leqq d$. The left member, after s differentiations, will consist of terms each having a factor of the form $k(k-1) \ldots (k-h)$, a factor which is a power of $e^y + z$ and a factor which is a power of e^y. The last occurs to power s and lower; the highest degree occurs in the term $k(k-1) \ldots (k-s+1)(e^y+z)^{k-s}e^{sy}$, and all the other coefficients have degree less than s in k. Now we set $y = \log x$, $z = 1 - x$, and obtain

$$k^s x^s + \text{terms of lower degree in } k \text{ and } x$$
$$= \sum_{m=0}^{k} \binom{k}{m} m^s x^m (1 - x)^{k-m} .$$

Next we divide both members by k^s, multiply by the coefficient of x^s in f, and sum for $s = 0, 1, \ldots, d$. The result is

$$f(x) - \sum_{j=1}^{d} p_j(x) k^{-j} = \sum_{m=0}^{k} \binom{k}{m} f(\tfrac{m}{k})\, x^m (1 - x)^{k-m} ,$$

where $p_1, \ldots, p_d$ are polynomials. This establishes the lemma.

We now return to the proof of the theorem, and set up a system satisfying postulates (22.1σ). For T we choose the closed half-line $[0,\infty]$. For f_1 and f_2 in F, we define $f_2 \gg f_1$ to mean that $f_2 - f_1$ is defined and has a positive lower bound on every interval $[0,t]$, t finite, and there exist real numbers $a \geqq 0$, b and $c > b$ such that $f_2(t) \geqq c \exp(-at) > b \exp(-at) \geqq f_1(t)$ for all large t. Now (1.3a,b) are obvious. With the notation of (1.3c'), there exist $a \geqq 0$, b, $c > b$ such that for large T, $f(t) \geqq c \exp(-at) > b \exp(-at) \geqq h(t)$, and numbers $a' \geqq 0$, b', $c' > b'$ such that for large t, $g(t) \geqq c' \exp(-at) > b' \exp(-at) \geqq k(t)$. The functions $b \exp(-at)$ and $b' \exp(-a't)$ have the same value at not more than a single point t unless they are identical; so we may assume $b \exp(-at) \geqq b' \exp(-a't)$ for all large t. Then for large t,

$$h(t) \vee k(t) \leqq b \exp(-at) \leqq c \exp(-at) \leqq f(t) \leqq f(t) \vee g(t) .$$

Since $h \vee k - f \vee g$ clearly has a positive lower bound on each interval $[0,t]$, $t > 0$, we have $f \vee g \gg h \vee k$. Likewise $f \wedge g \gg h \wedge k$, establishing (22.1c). We have assumed (22.1d).

The set E shall consist of all finite sums of exponentials

$$e(t) = c_1 \exp(-a_1 t) + \ldots + c_n \exp(-a_n t)$$

with $a_1, \ldots, a_n$ positive rational numbers. We map E into G by the elementary mapping I_0, where with the notation just used we define

$$I_0(e) = c_1 g(a_1) + \ldots + c_n g(a_n) .$$

Postulate (22.1i) is easily verified. If

$$e_1(t) = \sum_1^n c_i \exp(-a_i t) \quad \text{and} \quad e_2 = \sum_1^n c_i' \exp(-a_i' t) ,$$

and a_o is half of the least of the a_i and a_i', we define

$$e'' = \exp(-a_o t) + \sum_1^n |c_i| \exp(-a_i t) + \sum_1^n |c_i'| \exp(-a_i' t), \; e' = -e'' ,$$

and thus establish (22.1g), without the hypothesis that $I_o(e_1)$ and $I_o(e_2)$ have a common bound.

Let e_1, e_2 and e_3 be in E, and let f be a member of F such that $f \gg \text{mid}\,(e_1, e_2, e_3)$. For all large t, mid (e_1, e_2, e_3) coincides with one of the e_i, say e_1. Let e_1 be $c_1\exp(-a_1 t) + \ldots + c_n \exp(-a_n t)$, where $a_1 < a_2 \ldots < a_n$ and $c_1 \neq 0$. Then $\liminf f(t) \exp a_1 t \geqq c_1$. If equality held, there could exist no numbers $a > 0$, b, $c > b$ which could satisfy the requirements in the definition of $f \gg \text{mid}\,(e_1, e_2, e_3)$. Define $\tau = \exp(-a_1 t)$. Then $f(t(\tau))/\tau > c_1$ for τ near 0, so $f(t(\tau))/\tau - \text{mid}\,(e_1(t(\tau)), e_2(t(\tau)), e_3(t(\tau)))$ has a positive lower bound (say 4ϵ) for $0 < \tau \leqq 1$. Let $p(\tau)$ be a polynomial which approximates

$$\text{mid}\,(e_1(t(\tau)),\, e_2(t(\tau)),\, e_3(t(\tau))) + 2\epsilon$$

to within ϵ, and let $e(t) = p(\tau(t))$. Then e is in E, and $f \gg e \gg \text{mid}\,(e_1, e_2, e_3)$. This and its dual establish (22.1h).

Let e be a non-negative member of E. With the previous notation, let D be the least common denominator of the rational numbers $a_1, \ldots, a_n$, each in lowest terms. Then e is a polynomial $\{p(\tau) : 0 < \tau \leqq 1\}$ in the variable $\tau = \exp(-t/D)$. Let d be its degree, and let $p_1, \ldots, p_d$ be the polynomials in τ described in Lemma 30.2. For $k > d$, the Bernstein polynomials $B_k p$ satisfy

$$p(\tau) = B_k p(\tau) + \sum_{m=1}^d k^{-m} p_m(\tau) .$$

We transform back to the variable t and apply I_o. The left member is $I_o(e)$. The first term in the right is

$$B_k p(t(\tau)) = \sum_{m=o}^k e(\exp[-\tfrac{m}{kD}])\binom{k}{m}\exp(-\tfrac{mt}{D})\,[1 - \exp(-\tfrac{t}{D})]^{k-m} .$$

Each term $\exp(-mt/D)[1 - \exp(-t/D)]^{k-m}$ can be expanded by the binomial theorem; to the result we apply I_o, obtaining

$$g(\tfrac{m}{D}) - \binom{k-m}{1} g(\tfrac{[m+1]}{D}) + \binom{k-m}{2} g(\tfrac{[m+2]}{D}) - + \cdots + (-1)^{k-m} g(\tfrac{k}{D}) .$$

But this is $(-1)^{k-1} \Delta^{k-m}_{m/D}\, g(m/D)$, which by hypothesis is $\geqq \theta$. Since the coefficients $e(\exp[-m-kD]) \binom{k}{m}$ are non-negative, we have $I_0(B_k p(\tau(\cdot))) \geqq \theta$.

Each of the terms $k^{-m} p_m(\tau)$ transforms under $\tau = \tau(t)$ into $k^{-m} e_m$, where e_m is a fixed member of E. Hence by (7.4) $I_0(k^{-m} e_m)$ tends to θ. With the preceding proof, this shows

$$I_0(e) \geqq \sum_{m=1}^{d} k^{-m} I_0(e_m) ;$$

the right member tends to θ, so $I_0(e) \geqq 0$. Hence I_0 is isotone, and (22.1e) holds.

Finally, let S be a countable subset of E directed by $\ll$ having $\bigwedge S \leqq 0$ and such that $I_0(S)$ has a lower bound. Let e be a member of S, defined by $e(f) = c_1 \exp(-a_1 t) + \cdots + c_n \exp(-a_n t)$, where $a_1 < a_2 < \cdots < a_n$. If ϵ is positive, for all t greater than a certain t_0 we have $e(t) < \epsilon \exp(-a_1 t/2)$. On the compact set $[0, t_0]$ the members of S are directed by $\leqq$ and have limit $\leqq 0$, so they are eventually $\leqq \epsilon \exp(-a_1 t/2)$. Hence there exists a member e_1 of S such that $e_1(t) \leqq \epsilon \exp(-a_1 t/2)$ for all t in $[0,\infty)$, whence $I_0(e_1) \leqq \epsilon\, g(a_1/2)$. Consequently, $\bigwedge I_0(S) \leqq \epsilon\, g(a_1/2)$. (We use the restriction of countability of S to establish the existence of $\bigwedge I_0(S)$ and nowhere else). Since ϵ is arbitrary, $\bigwedge I_0(S) \leqq \theta$, and (22.1f) holds.

Now all the postulates (22.1σ) have been verified, so the integral I_0 can be used to define an integral I and a measure μ, and as in Sections 22 and 23 we have $I(f) = \int_0^\infty f(t)\mu(dt)$ for all summable functions f. Postulate (15.4) is easily verified, so that $I(e) = I_0(e)$ for all elements e of E. In particular, if e is defined by $e(t) = \exp(-at)$ with a rational and > 0, then

$$g(a) = I_0(e) = I(e) = \int_0^\infty \exp(-at)\, \mu(dt) .$$

Now let α be any positive real number, and let $a_1, a_2, \ldots$ be a sequence of positive rationals approaching α. The continuity of g is a straightforward consequence of its complete monotoneity; by the dominated convergence theorem we have

$$g(\alpha) = \lim_n g(a_n) = \lim_n \int_0^\infty \exp(-a_n t)\, \mu(dt)$$
$$= \int_0^\infty \exp(-\alpha t)\, \mu(dt) ,$$

and the proof is complete.

The analogue of Theorem 30.1 which deals with functions completely monotonic on the closed half-line $[0,\infty)$ is in fact a special case of (30.1). Its extension to n dimensions might have some interest. To avoid excessive notational complexity we limit ourselves to $n = 2$, but the methods extend without change to higher values of n.

Let $g = (g(s,t) : 0 \leqq s < \infty, 0 \leqq t < \infty)$ take values in the partially-ordered linear space G. For non-negative α and β and positive h and k, define

$$\Delta^1_{h,s} g(\alpha,\beta) = g(\alpha + h,\beta) - g(\alpha,\beta) ,$$

$$\Delta^1_{k,t} g(\alpha,\beta) = g(\alpha,\beta + k) - g(\alpha,\beta) ,$$

$$\Delta^{n+1}_{h,s} \Delta^m_{k,t} g(\alpha,\beta) = \Delta^1_{h,s}[\Delta^n_{h,s}\Delta^m_{k,t} g(\alpha,\beta)]$$

$$\Delta^n_{h,s} \Delta^{m+1}_{k,t} g(\alpha,\beta) = \Delta^1_{k,t}[\Delta^n_{h,s} \Delta^m_{k,t} g(\alpha,\beta)] .$$

We say that g is <u>completely monotone</u> if $(-1)^{m+n}\Delta^n_{h,s}\Delta^m_{k,t} g(\alpha,\beta) \geqq \theta$ for all non-negative integers m and n, all non-negative α and β, and all positive h and k. The theorem to be established is

(30.3) THEOREM. Let $g = \{g(s,t): 0 \leqq s < \infty, 0 \leqq t < \infty\}$ take on values in a Dedekind σ-complete normal partially-ordered linear system. If g is completely monotonic, there exists a countably additive non-negative bounded measure function μ, defined on a family of subsets of the quadrant $\{(s,t): 0 \leqq s < \infty, 0 \leqq t < \infty\}$ which includes all Borel sets, such that

$$g(\alpha,\beta) = \int_0^\infty \int_0^\infty \exp(-\alpha t - \beta s)\mu d(\alpha,\beta) ;$$

and conversely.

In proving this we use the Bernstein polynomials in two variables. If f is defined on the square $[0,0;1,1]$, and h and k are positive integers, we define

$$B_{h,k} f(u,v) = \sum_{m=0}^{h} \sum_{n=0}^{k} f(\tfrac{m}{h},\tfrac{n}{k}) \binom{h}{m} \binom{k}{n} u^m(1 - u)^{h-m} v^n(1 - v)^{k-n} .$$

It is clear that if $f(u,v) = f_1(u)f_2(v)$, then

$$B_{h,k}f(u,v) = (B_h f_1(u))(B_k f_2(v)) .$$

Lemma (30.2) has an analogue in n dimensions; for simplicity we state it for two dimensions only.

(30.4) **LEMMA**. If $f = \{f(u,v): 0 \leqq u \leqq 1,\ 0 \leqq v \leqq 1\}$ is a polynomial of degree d_1 in u and d_2 in v, there exist polynomials $p_{o,o}, \ldots, p_{d_1 d_2}$ of degree at most d_1 in u and d_2 in v, having $p_{o,o}$ identically 0, such that for all integers h,k with $h \geqq d_1$ and $k \geqq d_2$ and each (u,v) in the unit square

$$f(u,v) - B_{h,k}f(u,v) = \sum_{m=o}^{d_1} \sum_{n=o}^{d_2} h^{-m} k^{-n} p_{m,n}(u,v) .$$

Let us write $\Pi_{i,j}$ for the function $\{u^i v^i: 0 \leqq u \leqq 1,\ 0 \leqq v \leqq 1\}$. Then f can be written as

$$f(u,v) = \sum_{i=o}^{d_1} \sum_{i=o}^{d_2} c_{i,j} \Pi_{i,j}(u,v) .$$

Hence, using Lemma 30.2,

$$\begin{aligned} B_{k,k}f(u,v) &= \sum_{i=o}^{d_1} \sum_{j=o}^{d_2} c_{i,j} B_{h,k} \Pi_{i,j}(u,v) \\ &= \sum_{i=o}^{d_1} \sum_{j=o}^{d_2} c_{i,j}(B_h \Pi_{i,o}(u))(B_k \Pi_{o,j}(v)) \\ &= \sum_{i=o}^{d_1} \sum_{j=o}^{d_2} c_{ij}(\Pi_{i,o}(u) + h^{-1}p_{i,o,1}(u) + \ldots + h^{-1}p_{i,o,i}(u)) \\ &\qquad (\Pi_{o,j}(v) + k^{-1}p_{o,j,1}(v) + \ldots + k^{-j}p_{o,j,j}(v)) , \end{aligned}$$

where the $p_{i,j,l}$ are polynomials in u or in v independent of h and k, and of degree at most $i + j$. If we multiply out the parentheses, the term obtained by using the first term in each parenthesis is $\Sigma\, \Sigma\, c_{ij} \Pi_{i,o}(u) \Pi_{o,j}(v)$, which is $f(u,v)$. The other terms are each the product of factors $h^{-m} k^{-n}$ (m,n not both 0) by a polynomial independent of h and k, and of the degree described in the conclusion of the lemma. This completes the proof.

In analogy with the previous procedure, we define E to consist of all functions $e = (e(u,v): o \leqq u < \infty,\ o \leqq u < \infty)$ which are finite sums

$$e(u,v) = \sum_{i=0}^{d_1} \sum_{j=0}^{d_2} c_{ij}\exp[-(a_{ij}u + b_{ij}v)]$$

where a_{ij} and b_{ij} are non-negative rational numbers. Since the functions under investigation are defined on the closed quadrant, we can simplify the definition of $\gg$ somewhat; we say that $f_2 \gg f_1$ if $f_2 - f_1$ is defined and has a positive lower bound on every compact subset of the quadrant, and there exist numbers b and $c > b$ such that whenever $u + v$ is sufficiently large, $f_2(u,v) \geqq c > b \geqq f_1(u,v)$.

The details of the proof are so little changed, apart from the notational complexity, that there is no point in writing them out.

§ 31. SPECTRAL RESOLUTION OF REAL PARTIALLY ORDERED ALGEBRAS WITH UNIT

Let A be a partially ordered set linear system. Assume that a binary multiplication is defined on all of $A \times A$ such that the system is a linear system with binary multiplication as defined in (7.6). The binary multiplication will be assumed to be connected with the order-relation by the following conditions.

(31.1) POSTULATE.

(i) The product $(x.y : x \text{ in } A, y \text{ in } A)$ is o-continuous in each variable separately, and is commutative.

(ii) All squares $x^2 = x.x$ (hence all finite sums of squares) are $\geqq \theta$.

(iii) If $x \geqq 0$, there is a net $(y_\alpha : \alpha \text{ in } A)$ of (finite) sums of squares which is o-convergent to X.

In the presence of (31.1i), (31.1ii,iii) are equivalent to saying that the set of elements $\geqq \theta$ is the smallest o-closed set containing all sums of squares of elements of A.

Because of the commutativity of multiplication, the product of squares is a square, so that the product of sums of squares is a sum of squares. If x and y are $\geqq \theta$, there exist nets $(x_\alpha : \alpha \text{ in } A)$ and $(y_\beta : \beta \text{ in } B)$ such that each x_α and y_β is a sum of squares and $\text{o-lim}\, x_\alpha = x$ and $\text{o-lim}_\beta\, y_\beta = y$. Then for each β in B,

$$xy_\beta = \text{o-lim}_\alpha\, x_\alpha y_\beta \geqq \theta ,$$

$$xy = \text{o-lim}_\beta\, xy_\beta \geqq \theta ,$$

proving that if x and y are $\geqq \theta$ then $xy \geqq \theta$. By (5.5), the function $(xy : x \text{ in } A, y \text{ in } A, x \geqq \theta, y \geqq \theta)$ is smoothly rising. Thus by definition (7.8iii) the system A, with its three operations, is a partially ordered algebra.

(31.2) DEFINITION. A subset $B_o = \{x_\beta : \beta \text{ in } B\}$ is a σo-basis for A if no Dedekind σ-closed proper subset of F contains B_o and is an algebra embedded in A.

(31.3) THEOREM. Let A be a normal Dedekind-closed commutative partially ordered algebra satisfying postulate (31.1), and let $B_o = \{x_\beta : \beta \text{ in } B\}$ be a σo-basis for A. Assume that A contains a unit u, and that for each x in A there exist real numbers a, b such that $au \leqq x \leqq bu$. Let $T = \underset{\beta \text{ in } B}{\times} J_\beta$ be the cartesian product of closed intervals of real numbers, one for each β in B, such that if J_β is $[a_\beta, b_\beta]$ the basis element x_β satisfies $a_\beta u \leqq x_\beta \leqq b_\beta u$. Let F be the set of all extended-real-valued functions on T. Then there exists a mapping I of a lattice $F_{sum} \subset F$ onto the algebra A with the following properties.

(i) If f_1 and f_2 are in F_{sum} and are finite valued, and a_1 and a_2 are real numbers, then $a_1 f_1 + a_2 f_2$ is in F_{sum}, and $I(a_1 f_1 + a_2 f_2) = a_1 I(f_1) + a_2 I(f_2)$.

(ii) If a sequence $(f_n : n = 1, 2, 3, \ldots)$ of functions in F_{sum} is uniformly bounded and converges everywhere in T to a limit f_o, then f_o is in F_{sum}, and $I(f_o) = o\text{-lim } I(f_n)$.

(iii) F_{sum} contains all bounded Borel-measurable functions.

(iv) To each x in A corresponds at least one bounded function of Baire f such that $I(f) = x$.

(v) If f_1 and f_2 are finite-valued and in F_{sum}, so is $f_1 f_2$, and $I(f_1 f_2) = I(f_1) I(f_2)$.

(vi) If β is in B, and p_β is the function which for each $t = (t_{\beta'} : \beta' \text{ in } B)$ is the β-component t_β of t, then $I(p_\beta) = x_\beta$.

(vii) The sets measurable with respect to I include all Borel sets. If M is measurable, its measure mM is an element of A satisfying $mM \geqq 0$, $mM = (mM)^2$. The measure of the whole space T is $mT = u$. If M_1 and M_2 are measurable sets, $m(M_1 \cap M_2) = (mM_1)\cdot(mM_2)$.

(viii) If f is in F_{sum} and $I(f) \geqq \theta$, then $m\{t:f(t) < 0\} = \theta$. If f is in F_{sum} and is bounded below, and $m\{t:f(t) < 0\} = \theta$, then $I(f) \geqq \theta$. If f_1 and f_2 are bounded and equivalent, and f_1 is in F_{sum}, so is f_2, and $I(f_2) = I(f_1)$. If f_1 and f_2 are finite-valued and in F_{sum} and $I(f_1) = I(f_2)$, then $m\{t:f_1(t) \neq f_2(t)\} = \theta$.

(ix) There is a one-to-one correspondence between the elements of A and the equivalence-classes of finite-valued summable functions; each such equivalence-class contains at least one bounded function of Baire.

If we define $f_1 \gg f_1$ to mean that $f_2(t) - f_1(t)$ is defined for all t in T and has a positive lower bound, we verify readily that (22.1c) holds; (22.1a,b,d) are obvious. The set E shall consist of all functions e which are polynomials in finitely many of the components t_β of t. Let e depend only on the components t_β with β in a finite subset $B' = \{x_{\beta_1}, \ldots, x_{\beta_n}\}$ of B; if e is in E, it has the form

$$\sum c_{i_1}, \ldots, {}_{i_n}(t_{\beta_1})^{i_1} \ldots (t_{\beta_n})^{i_n},$$

where $i_1, \ldots, i_n$ each range over some finite set of non-negative integers. Corresponding to this e we define

$$I_o(e) = \sum c_{i_1}, \ldots, {}_{i_n}(x_{\beta_1})^{i_1} \ldots (x_{\beta_n})^{i_n}.$$

By Tychonoff's theorem, T is compact (in the usual product-topology), and each element of E is continuous on T, so (22.1g) is obvious. Also, if e_1, e_2 and e_3 are in E, mid (e_1,e_2,e_3) is continuous on T and depends on only finitely many of the components, say $\{t_{\beta'}:\beta'$ in $B'\}$. If $f \gg$ mid (e_1,e_2,e_3), let 4ϵ be the lower bound of their difference, and let e be a polynomial in $\{t_{\beta'}:\beta'$ in $B'\}$ which approximates mid $(e_1,e_2,e_3) + 2\epsilon$ to within ϵ. This satisfies the requirements in (22.1h). Moreover, (22.1i) is obviously satisfied.

Let e be a non-negative member of E; it depends on finitely many components of t, which for notational simplicity we designate by $t_1, \ldots, t_n$. The corresponding intervals in the cartesian product $T = \times_\beta J_\beta$ are $J_i = [a_i,b_i]$, $i = 1, \ldots, n$. We map each J_i linearly on $[0,1]$ by the mapping $t_i' = (t - a_i)/(b_i - a_i)$, obtaining a polynomial $e' = (e'(t'): 0 \leqq t_i' \leqq 1, i = 1, \ldots, n)$ non-negative on its domain. The corresponding Bernstein polynomial of degrees $k_1, \ldots, k_n$ in $t_1', \ldots, t'_n$ respectively is

$$B_{k_1,\ldots k_n}e'(t_1', \ldots, t_n') = \sum_{m_1=0}^{k_1} \cdots \sum_{m_n=0}^{k_n} e'\left(\frac{m_1}{k_1}, \ldots, \frac{m_n}{k_n}\right)\binom{k_1}{m_1} \cdots$$

$$\binom{k_n}{m_n}(t_1')^{m_1}(1 - t_1')^{k_1-m_1} \cdots (t_n')^{m_n}(1 - t_n')^{k_n-m_n} .$$

According to Lemma (30.4), extended to n dimensions, if $d_1, \ldots, d_n$ are the degrees of e' in $t_1', \ldots, t_n'$ respectively, there exist polynomials $p'_{j_1,\ldots,j_n}$ $(j_i = 1, \ldots, d_i,\ i = 1, \ldots, n)$ such that for all large $k_1, \ldots k_n$

$$e'(t_1',\ldots,t_n') = B_{k_1,\ldots,k_n}e'(t_1',\ldots,t_n')$$

$$+ \sum_{j_1=0}^{d_1}\cdots\sum_{j_n=0}^{d_n} (k_1)^{-j_1}\cdots(k_n)^{-j_n}\, p_{j_1,\ldots,j_n}(t'_1,\ldots,t_n') ,$$

where $p_{0,0,\ldots,0}$ is identically zero. We now transform back to the variables t_β and apply I_0. The left member is $I_0(e)$. Each term in the Bernstein polynomial consists of a non-negative real multiple of factors of the form $(t_i')^{m_i}(1 - t_i')^{k_i-m_i}$, which transforms into a positive multiple of $(t_i - a_i)^{m_i}(b_i - t_i)^{k_i-m_i}$. In applying I_0 each such factor is replaced by $(x_i - a_iu)^{m_i}(b_iu - x_i)^{k_i-m_i}$, which is $\geqq \theta$. Hence the result of applying I_0 to the Bernstein polynomial is $\geqq \theta$. The result of applying I_0 to the polynomial $p_{j_1,\ldots,j_n}$ is some fixed element $y_{j_1,\ldots,j_n}$ of A. Hence

$$I_0(e) \geqq \sum_{j_1=0}^{d_1} \cdots \sum_{j_n=0}^{d_n} (k_1)^{-j_1} \cdots (k_n)^{-j_n}\, y_{j_1,\ldots,j_n} .$$

But since $y_{0,0,\ldots 0} = 0$, this approaches θ as $k_1, \ldots, k_n$ increase (cf. (7.4)), so $I_0(e) \geqq \theta$, and (22.1e) is established.

The proof of (22.1f) is essentially the same as in, say, Section 26, so all of postulate (22.1) holds. Hence we can define an integral I on a subset F_{sum} of F, having all the properties established in Section 22 and earlier sections. In particular, (15.4) holds for each e_0 in E; we need only take $e_i' = e_0 - 1/i$, $e_i'' = e_0$, $e_i''' = e_0 + 1/i$, $i = 1, 2, 3, \dots$. So $I(e) = I_0(e)$ for every e in E, which implies conclusion (vi). Conclusion (i) follows from (22.4), and (ii) from (15.1).

It is not difficult to show that a function is a U-function if and only if it is bounded below and lower semi-continuous. In particular, the characteristic function of an open set is a U-function, so by (17.5) is a measurable function. Thus all open sets are measurable. In particular, with $e \equiv 1$, we obtain $mT = I_0(1) = u$. Since differences, countable unions and countable intersections of measurable sets are measurable, all Borel sets are measurable. Consequently all Borel-measurable functions are measurable, which implies (iii).

Conclusion (v) follows at once from (19.5). In particular, if f is the characteristic function of a measurable set M, then $f = f^2$, so by (v) $I(f) = [I(f)]^2$, that is, $mM = (mM)^2 \geqq \theta$. If f_1 and f_2 are the characteristic functions of measurable sets M_1 and M_2 respectively, f_1f_2 is the characteristic function of $M_1 \cap M_2$, so $(mM_1)(mM_2) = m(M_1 \cap M_2)$, and (vii) is established.

Suppose that f is summable (hence measurable) and that $I(f) \geqq \theta$, but $m\{t: f(t) < 0\} \neq 0$. Since $m\{t: f(t) < 0\} = \text{o-lim}_{n \to \infty} m\{t: f(t) \leqq -1/n\}$, there exists a positive integer n such that the set $M = \{t: f(t) \leqq -1/n\}$ has measure $mM \neq \theta$. Then on the one hand $I(f \cdot \chi_M) = I(f)I(\chi_M) = I(f)mM \geqq \theta$. On the other hand, $f \cdot \chi_M \leqq -\chi_M/n$, so

$$I(f \cdot \chi_M) \leqq -n^{-1}I(\chi_M) = -n^{-1}mM .$$

Combining these estimates yields $-n^{-1}mM \geqq \theta$, whence $mM \leqq \theta$. But by (vii) $mM \geqq \theta$, so $mM = \theta$, which is a contradiction. This establishes the first sentence in (viii).

If f is summable and has a lower bound -b, and the set $N = \{t: f(t) < 0\}$ has $mN = \theta$, then the function $f + b\chi_N$ is $\geqq 0$ on T, so $\theta \leqq I(f + b\chi_N) = I(f) + bI(\chi_N) = I(f) + bmN = I(f)$, establishing the second sentence. If f_1 and f_2 are bounded and equivalent, and f_1 is summable, so is f_2 by (22.15). The sets $\{t: f_2(t) - f_1(t) < 0\}$ and $\{t: f_1(t) - f_2(t) < 0\}$ both have measure θ, so by the proof just completed we have $I(f_2 - f_1) \geqq \theta$ and $I(f_1 - f_2) \geqq \theta$, establishing the third sentence of (viii).

Suppose that f_1 and f_2 are summable and finite-valued, and $I(f_1) = I(f_2)$. By the first sentence of (viii), both $m\{t: f_1(t) - f_2(t) < 0\}$ and $m\{t: f_2(t) - f_1(t) < 0\}$ are equal to θ. Hence the union of these sets, which is $\{t: f_1(t) \neq f_2(t)\}$, has measure θ, establishing the last sentence of (viii).

Returning to (iv), we let K be the subset of A consisting of those elements x to which there corresponds a bounded function of Baire f such that $I(f) = x$. All elements of the σo-basis B_o are in K, since by (vi) each corresponds to a monomial. We now prove that K is Dedekind σ-closed. Let S be a countable subset of K directed by $\geqq$ and having an upper bound; it then has a supremum $\bigvee S$ in A. By (2.4), we may suppose that S is an isotone sequence $x_1 \leqq x_2 \leqq \cdots$ converging to $\bigvee S$. To each x_n corresponds a bounded function of Baire f_n' such that $I(f_n') = x_n$. Let a be a number such that $\bigvee S \leqq au$. Since $x_n \leqq cu$ for each n, by (viii) we have $f_n' \leqq c$ except on a set of measure θ. Hence f_n' is equivalent to $f_n' \wedge c$, which is a Baire function; and by (viii) $I(f_n' \wedge c) = x_n$. That is, we may as well suppose that all the f_n' satisfy $f_n' \leqq c$ to begin with. Next we define inductively $f_1 = f_1', \ldots, f_n = f_n' \vee f_{n-1}$. There are clearly Baire functions. The relation $I(f_n) = x_n$ holds for $n = 1$ by definition. If it holds for $n = k - 1$, then $I(f_k') = x_k \geqq x_{k-1} = I(f_{k-1})$, so by (viii) $f_k'(t) \geqq f_{k-1}(t)$ except on a set of measure θ, and f_k' is equivalent to $f_k' \vee f_{k-1}$, which is f_k. Then by (viii) $I(f_k) = I(f_k') = x_k$, and so by induction we have $I(f_n) = x_n$ for all n. Since $f_1 \leqq f_2 \leqq \cdots \leqq c$, they have a limit f_o which is also a Baire function and is $\leqq c$. By (ii), $I(f_o) = \text{o-lim}\, I(f_n) = \text{o-lim}\, x_n = \bigvee S$. Hence $\bigvee S$ is also in K. In a like manner, if S is a countable subset of K directed by $\leqq$ and having an infimum in A, $\bigwedge S$ is in K. Hence K is Dedekind σ-complete. It is obviously closed under addition, scalar multiplication and binary multiplication, by (i) and (iv), so it is an algebra embedded in A. But by the definition (31.2) of a σo-basis, K cannot be a proper subset of A. So $K = A$; and (iv) is established.

Statement (ix) is merely a collection of assertions already established in (iv) and (viii).

As a particular case, let B be a bounded hermitian operator on a Hilbert space H. The set $cc(B)$ defined after (29.3) is a Dedekind-closed commutative sub-algebra of the algebra of all linear operators on H, and it contains B. Let A be the intersection of all Dedekind-closed commutative sub-algebras of the algebra of linear operators which contain B as a member. This is again such an algebra. By (29.3), (31.ii) is satisfied; (31.iii) is obvious, and (31.iiii) holds as a by-product of the proof of (28.6). By definition of A, the single element B is a σo-basis for A. Then the hypotheses of (31.3) hold, T being a single interval of real numbers; and (31.3) provides the spectral resolution of B discussed in Section 29.

§ 32. SPECTRAL RESOLUTION OF COMPLEX PARTIALLY ORDERED *-ALGEBRAS WITH UNIT

As our final application we consider a commutative algebra in which the scalar coefficient - field consists of the complex numbers. We assume that there is a mapping $*$ of A into itself such that for all x and y in A and all complex numbers c we have

$$(x^*)^* = x ,$$

$$(x + y)^* = x^* + y^* ,$$

$$(xy)^* = y^*x^* ,$$

$$(cx)^* = \bar{c}\, x^* .$$

The element x^* is the "adjoint" of x; if $x^* = x$, x is self-adjoint.

For each x in A we define the "symmetric" and "skew" parts of x by the equations

$$\text{Sy } x = \frac{(x + x^*)}{2} .$$

$$\text{Sk } x = \frac{(x - x^*)}{2i} .$$

Each of these is self-adjoint, and

$$x = \text{Sy } x + i\text{Sk } x .$$

The self-adjoint elements of A constitute an algebra A_{sa} over the field of real numbers. We suppose that this real algebra is partially ordered by a relation $\geqq$ with which it is a partially ordered linear system and (31.1) is satisfied. Then we can partially order A by defining $x_2 \geqq x_1$ to mean $\text{Sy } x_2 \geqq \text{Sy } x_1$, $\text{Sk } x_2 \geqq \text{Sk } x_1$. It follows readily that if $(x_\beta : \beta$ in $B)$ is a net of elements of A and x_0 is in A, the relation

$$\text{o-lim}_\beta\, x_\beta = x_0$$

is equivalent to the two relations

$$\text{o-lim}_\beta \text{ Sy } x_\beta = \text{Sy } x_0 ,$$

$$\text{o-lim}_\beta \text{ Sk } x_\beta = \text{Sk } x_0 .$$

If A has a unit u, then $u^* = uu^*$ and $u = (u^*)^* = (u^*)^*u^* = uu^* = u^*$, so u is self-adjoint.

A sub-algebra A' of A is called <u>self-adjoint</u> if for each x in A', x^* is also in A'.

A subset $B_0 = \{x_\beta : \beta \text{ in } B\}$ of A is a (complex) σo-basis for the *-algebra A if no proper self-adjoint Dedekind σ-closed sub-algebra of A contains B_0. It is evident that B_0 is a (complex) σo-basis for A if and only if the elements Sy x_β and Sk x_β (x_β in B_0) also form a σo-basis; and this in turn is true if and only if the elements Sy x_β and Sk x_β (x_β in B_0) form a (real) σo-basis for the (real) algebra A_{sa} of self-adjoint elements of A.

(32.1) THEOREM. Let A be a normal Dedekind-closed commutative partially ordered *-algebra whose self-adjoint elements satisfy (31.1), and let $B_0 = \{x_\beta : \beta \text{ in } B\}$ be a (complex) σo-basis for A. Assume that A contains a unit u, and that for each x in A there exist real numbers a,b such that $au \leqq \text{Sy } x \leqq bu$ and $au \leqq \text{Sk } x \leqq bu$. Let $T = \mathop{\times}\limits_{\beta \text{ in } B} J_\beta$ be the cartesian product of closed intervals of complex numbers, one for each β in B, such that if J_β is the interval

$$\{z: a'_\beta \leqq \text{Re } z \leqq b'_\beta,\ a''_\beta \leqq \text{Im } z \leqq b''_\beta\}$$

the basis element x_β satisfies the inequalities

$$a'_\beta u \leqq \text{Sy } x_\beta \leqq b'_\beta u,\ a''_\beta u \leqq \text{Sk } x_\beta \leqq b''_\beta u .$$

Let F be the set of all extended-complex-valued functions on T (that is, all functions of the form $f' + if''$, f' and f'' extended-real-valued). Then there exists a mapping I of a lattice $F_{sum} \subset F$ onto the algebra A with the following properties.

(1) If f_1 and f_2 are in F_{sum} and are finite-valued, and c_1 and c_2 are complex numbers, then $c_1 f_1 + c_2 f_2$ is in F_{sum}, and $I(c_1 f_1 + c_2 f_2) = c_1 I(f_1) +$

$c_2I(f_2)$; and $\overline{f}_1$ is in F_{sum}, and $I(\overline{f}_1) = [I(f_1)]^*$.

(ii) If a sequence $(f_n: n = 1, 2, \ldots)$ of functions in F_{sum} is uniformly bounded and converges everywhere in T to a limit f_o, then f_o is in F_{sum}, and $I(f_o) = \text{o-}\lim_{n\to\infty} I(f_n)$.

(iii) F_{sum} contains all bounded Borel-measurable functions.

(iv) To each x in A corresponds at least one bounded function of Baire such that $I(f) = x$.

(v) If f_1 and f_2 are finite-valued and in F_{sum}, f_1f_2 is also in F_{sum}, and $I(f_1f_2) = I(f_1)\, I(f_2)$.

(vi) If β is in B, and p_β is the function which for each t in T is the β-component t_β of t, then $I(p_\beta) = x_\beta$ and $I(\overline{p}_\beta) = x_\beta^*$.

(vii), (viii) (Identical with (31.3vii,viii) respectively.)

With the notation of the theorem, let J'_β be the interval $\{t': a'_\beta \leqq t' \leqq b'_\beta\}$ and J''_β the interval $\{t'': a''_\beta \leqq t'' \leqq b''_\beta\}$. Then $T = \underset{\beta \text{ in } B}{\times} (J'_\beta \times J''_\beta)$. By the sentence just before (32.1), the elements $\text{Sy}\, x_\beta$, $\text{Sk}\, x_\beta$, β in B form a (real) σo-basis for the algebra A_{sa} of self-adjoint elements of A. We let the intervals J'_β, J''_β correspond respectively to $\text{Sy}\, x_\beta$ and $\text{Sk}\, x_\beta$ for each β in B. Now we can apply (31.3) and obtain a mapping I of a lattice F'_{sum} of extended-real-valued functions on T having all the properties specified in the conclusions of (31.3); in particular, (vii) and (viii) of the present theorem hold. If f' and f'' are in F'_{sum}, we define $I(f' + if'')$ to be $I(f') + iI(f'')$. Conclusions (ii) and (iii) are immediate, and so is the linearity in (i). If $f_1 = f' + if''$, with f' and f'' real and summable, $I(f')$ and $I(f'')$ are self-adjoint (since I maps F'_{sum} onto A_{sa}), and so $I(\overline{f_1}) = I(f') - iI(f'') = I(f')^* + [iI(f'')]^* = [I(f_1)]^*$, completing the proof of (i). To establish (v) we need only express f_1 and f_2 in terms of their real and imaginary parts and use (31.3v). If x is in A, $\text{Sy}\, x$ and $\text{Sk}\, x$ are in A_{sa}, so by (31.3iv) there exist bounded functions of Baire f' and f'' such that $\text{Sy}\, x = I(f')$ and $\text{Sk}\, x = I(f'')$. Then $x = \text{Sy}\, x + i\text{Sk}\, x = I(f' + if'')$, and $f = f' + if''$ serves in conclusion (iv).

Let p'_β be the function which at t has for its value the real part t'_β of the component t_β of t in J, and similarly let p''_β have the value t''_β. Then by (31.3vi) $I(p'_\beta) = \text{Sy } x_\beta$, $I(p''_\beta) = \text{Sk } x_\beta$. So $I(p_\beta) = I(p'_\beta + ip''_\beta) = \text{Sy } x_\beta + i \text{ Sk } x_\beta = x_\beta$. The remaining conclusion of (vi) follows from (i).

BIBLIOGRAPHY

BIRKHOFF, GARRETT

[1] "Lattice theory," American Mathematical Society Colloquium Publications XXV (1948).

BOCHNER, S.

[1] "Integration von Funktionen, deren Werte die Elemente eines Vektorraumes sind," Fundamenta Math. 20 (1933), 262-276.

[2] "Completely monotone functions in partially ordered spaces," Duke Math. Journal 9 (1942), 519-526.

BOCHNER, S. and KY FAN

[1] "Distributive order-preserving operations in partially ordered vector sets," Annals of Mathematics, series 2 48 (1947), 168-179.

DANIELL, P. J.

[1] "A general form of integral," Annals of Mathematics, series 2 19 (1917-18), 281-288.

FELL, J. M. G. and KELLEY, J. L.

[1] "An algebra of unbounded operators," Proc. Nat. Acad. Sci. 38 (1952), 592-598.

FREUDENTHAL, HANS

[1] "Teilweise geordnete Moduln," Proceedings of Section of Science, K. Akad. van Wetenschappen te Amsterdam 39 (1936), 641-651.

KANTOROVITCH, L.

[1] "Lineare halbgeordnete Räume," Mat. Sbornik n.s. 2 44 (1937), 121-168.

KELLEY, J. L.

[1] "Convergence in topology," Duke Math. Journal 17 (1950), 277-283.

LORCH, E. R.

[1] "Functions of self-adjoint operators in Hilbert space," Acta Litt. ac Sci. Szeged 7 (1934), 136-146.

McSHANE, E. J.

[1] Integration, Princeton University Press (1949).

McSHANE, E. J. and BOTTS, T. A.

[1] "A modified Riemann-Stieltjes integral," Duke Math. Journal 19 (1952), 293-302.

NAGY, B. v. S.

[1] Spektraldarstellung linearer transformationen des Hilbertschen Raumes, Springer-Verlag, Berlin (1942).

NAKANO, HIDEGORÔ

[1] Modulared Semi-ordered Linear Spaces, Maruzen Co. Ltd., Nihonbashi, Tokyo.

[2] Modern Spectral Theory, Maruzen Co. Ltd., Nihonbashi, Tokyo.

von NEUMANN, JOHN

[1] "Allgemeine Eigenwerttheorie Hermitescher Funktionaloperatoren," Math. Annalen 102 (1929), 49-131.

STONE, M. H.

[1] "Notes on integration, I, II, III, IV," Proc. Nat. Acad. Sci. 34 (1948), 336-342, 447-455, 483-490; 35 (1949), 50-58.

[2] "The generalized Weierstrass approximation theorem," Math. Mag. 21 (1948), 167-184.

[3] "A general theory of spectra," Proc. Nat. Acad. Sci. 26 (1940), 280-283 and 27 (1941), 83-87.

PRINCETON MATHEMATICAL SERIES

Edited by Marston Morse and A. W. Tucker

9 780691 095820

Ingram Content Group UK Ltd.
Milton Keynes UK
UKHW032005260523
422428UK00001B/11

9 780691 095820